99-WSM-3

TECHNICAL APPROACHES FOR SETTING SITE-SPECIFIC NUTRIENT CRITERIA

by:

William J. Warren-Hicks, Ph.D.
Benjamin R. Parkhurst, Ph.D.
Steven M. Bartell, Ph.D.
The Cadmus Group, Inc.

Miles M. Smart, Ph.D.
Smart & Associates

2005

WERF

The Water Environment Research Foundation, a not-for-profit organization, funds and manages water quality research for its subscribers through a diverse public-private partnership between municipal utilities, corporations, academia, industry, and the federal government. WERF subscribers include municipal and regional water and wastewater utilities, industrial corporations, environmental engineering firms, and others that share a commitment to cost-effective water quality solutions. WERF is dedicated to advancing science and technology addressing water quality issues as they impact water resources, the atmosphere, the lands, and quality of life.

For more information, contact:
Water Environment Research Foundation
635 Slaters Lane, Suite 300
Alexandria, VA 22314-1177
Tel: (703) 684-2470
Fax: (703) 299-0742
www.werf.org
werf@werf.org

This report was co-published by the following organizations. For non-subscriber sales information, contact:

Water Environment Federation
601 Wythe Street
Alexandria, VA 22314-1994
Tel: (800) 666-0206
Tel: (703) 684-2452
Fax: (703) 684-2492
www.wef.org
pubs@wef.org

IWA Publishing
Alliance House, 12 Caxton Street
London SW1H 0QS, United Kingdom
Tel: +44 (0) 20 7654 5500
Fax: +44 (0) 20 7654 5555
www.iwapublishing.com
publications@iwap.co.uk

Library of Congress Catalog Card Number: 2003114927

WEF ISBN: 1-57278-230-7 IWAP ISBN: 1-84339-639-4

Cadmus Group, Inc., Smart & Associates

The research on which this report is based was funded, in part, by the United States Environmental Protection Agency through Cooperative Agreement No. CR-827345 with the Water Environment Research Foundation (WERF). Unless an EPA logo appears on the cover, this report is a publication of WERF, not EPA. Funds awarded under the Cooperative Agreement cited above were not used for editorial services, reproduction, printing, or distribution.

ACKNOWLEDGMENTS

This research was sponsored by the Water Environment Research Foundation. The principal investors appreciated the useful participation of the project team in the design and completion of this project.

Report Preparation

Principal Investigators:
William J. Warren-Hicks, Ph.D.
Benjamin R. Parkhurst, Ph.D.
Steven M. Bartell, Ph.D.
Cadmus Group, Inc.

Miles M. Smart, Ph.D.
Smart & Associates

Project Subcommittee:

David Flemer, Ph.D.
U.S. Environmental Proctection Agency

Lance Butler
Philadelphia Water Department

Norman E. Leblanc
Hampton Roads Sanitation District

James J. Sartoris
U.S. Geological Survey, Biological Resources Division

Paul E. Stacey
State of Connecticut, Bureau of Water Management

Water Environment Research Foundation Staff

Director of Research: Daniel M. Woltering, Ph.D.
Project Manager: Margaret Stewart

ABSTRACT AND BENEFITS

The purpose of this project was to develop a methodology for deriving site-specific nutrient criteria (SSNC) for surface waters, including streams and rivers, lakes and reservoirs, and coastal estuaries. The methodology was developed to extend the United States Environmental Protection Agency's regional nutrient criteria for localized conditions characterized by particular desired water quality requirements or designated uses. The proposed SSNC methodology provides local stakeholders with a recipe for estimating nutrient criteria consistent with site-specific water quality management goals and objectives.

The SSNC methodology prescribes a three-tiered or sequential approach for defining concentrations of acceptable nutrients in relation to management goals and objectives. Each tier requires successively more site-specific data and information and also develops increasingly quantitative and technologically more detailed relationships between nutrients and stated water quality measurements (chlorophyll *a*, Secchi depth, dissolved oxygen). The SSNC process can be initiated at any tier, although most applications will likely progress from Tier 1. The derivation of Tier 1 SSNC relies extensively on existing data and regional nutrient criteria. Tier 2 adds additional, more site-specific data and estimates SSNC on the basis of statistical relationships between nutrients and the selected water quality parameters of interest. Tier 3 extends Tier 2 through the development of additional site-specific data and the application of site-specific, process-level water quality models to estimate the SSNC. Follow-up monitoring is a key component of all three tiers for assessing the effectiveness of the SSNC in achieving the desired water quality characteristics and making subsequent decisions about continued implementation or modification of the SSNC.

Benefits:

- SSNC can serve as effective alternatives to regional criteria, which may fail to achieve or sustain locally desired water quality conditions.
- The proposed methodology prescribes an efficient and economical approach for achieving site-specific water quality objectives.
- The methodology develops SSNC on the basis of process-level understanding of relationships between nutrients and water quality objectives.
- The tiered approach permits a sequential, increasingly detailed and sophisticated analysis of relations between nutrients and desired water quality conditions.
- The results of the tiered SSNC methodology provide direct inputs to localized management and decision-making processes.

Keywords: Nutrient criteria, designated uses, desired water quality conditions, nitrogen, phosphorus, statistical nutrient models, mechanistic nutrient models.

TABLE OF CONTENTS

LIST OF TABLES

LIST OF FIGURES

LIST OF ACRONYMS

AF	anoxic factor
AHOD	areal hypolimnetic oxygen depletion rate
AQUATOX	aquatic ecosystem model for assessing risks from chemicals
BASINS	better assessment science integrating point and nonpoint sources
BMPs	best management practices
CASM	comprehensive aquatic system model
CCC	criterion continuous concentration
CFDs	cumulative frequency distributions
Chl *a*	chlorophyll *a*
CMC	criterion maximum concentration
DIN	dissolved inorganic nitrogen
DO	dissolved oxygen
GEM	general ecosystem model
HSPF	Hydrological Simulation Program
IBP	International Biological Program
IFEM	integrated fates and effects model
N	nitrogen
NGO	nongovernmental organization
NPDES	National Pollutant Discharge Elimination System
NSW	nutrient sensitive water
P	phosphorus
QUAL2E	enhanced streamwater quality model
SAV	submerged aquatic vegetation
Si	silica
SRP	soluble reactive phosphorus
SSNC	site-specific nutrient criteria
SWMM	stormwater management model
TIN	total inorganic nitrogen
TMDL	total maximum daily load
TN	total nitrogen
TP	total phosphorus
TSS	total suspended solids
U.S. EPA	United States Environmental Protection Agency
USGS	United States Geological Survey
VSS	volatile suspended solids
WASP	Water Quality Analysis Simulation Program
WERF	Water Environment Research Foundation
WQO	water quality objectives

EXECUTIVE SUMMARY

By 2004, if states and Native American tribes have not either adopted the U.S. EPA's regional nutrient criteria and standards or have prepared a plan to develop nutrient criteria, the U.S. EPA may promulgate criteria for them based on regional analyses and recommendations. Once the U.S. EPA publishes regional nutrient criteria, states, tribes, or other entities may propose their own site-specific nutrient criteria (SSNC) if they believe that regional criteria are not appropriate for particular surface waters of interest.

In the current U.S. EPA approach, no causal relationships between nutrients and response variables, such as chlorophyll *a* (Chl *a*) are determined. Consequently, in waters where such relationships are weak or may not exist because of confounding factors, management efforts to reduce nutrient concentrations to attain the ecoregional criteria may not produce predicted improvements in Chl *a* concentrations or other response variables. Also, relationships between nutrients and designated uses, or other water quality objectives, are not inherently determined in the ecoregional approach; consequently, efforts to reduce nutrient concentrations may not lead to the attainment of desired designated uses or other nutrient-affected water quality objectives.

This document provides detailed methods for deriving scientifically defensible SSNC and standards. The document is intended for use by 1) state and tribal regulatory agencies with the authority to derive criteria and standards; 2) regulated entities affected by the criteria, such as dischargers with National Pollutant Discharge Elimination System (NPDES) permits and those potentially affected by nonpoint source controls; and 3) other interested parties. The methods presented here for developing SSNC are meant to complement U.S. EPA's guidance for developing ecoregional nutrient criteria.

A three-tiered approach is proposed for developing SSNC. This approach would be implemented when the potential need for such criteria is determined based on evaluation of whether 1) the regional nutrient criteria are under-protective or are over-protective of uses and 2) designated uses and water quality objectives are either being attained or not being attained.. The focus of these site-specific methods is on the establishment of quantitative relationships among nutrients, response, and designated uses. These relationships are then used to develop site-specific nutrient criteria.

The tiered approach consists of a sequence of increasingly refined and detailed site-specific methods for deriving nutrient criteria. The demand for site-specific information also increases with each successive tier. In some parts of the U.S., nutrient data are available in a readily useable format. Examples of good lake survey information databases exist in Iowa, Texas, Minnesota, Missouri, Colorado, and Florida. Compilations of lake and reservoir data are available for some regions. In many regions, however, sufficient water quality information is available only for a small number of economically important waterbodies. Thus, characterizing site-specific water quality from existing data may prove difficult.

The tiers differ in the underlying data and analytical methods used to develop the criteria. Although the sequence involves increasingly detailed and data-intensive procedures, the process may be initiated at any of the three tiers. The main objective of each tier is to develop site-specific criteria.

Tier 1 methods for developing SSNC are based on the U.S. EPA's methods for developing regional criteria. The Tier I methods differ from the EPA ecoregional approach by focusing on data from the waterbody of concern and developing nutrient and response variable relationships, with existing data used as much as possible. Cumulative distributions of nutrient data for the waterbody are used to identify percentiles, such as the lower 25th percentiles of total nitrogen or total phosphorus, which are used as the proposed SSNC. The cause-and-effect relationships between nutrients and response variables are evaluated to determine the likely success of the proposed criteria for attaining designated uses and other water quality objectives. The Tier 1 assessment concludes with the development of a longer term monitoring plan and continued evaluation of the SSNC performance.

A Tier 2 analysis would be implemented if the Tier 1 SSNC were determined to be either underprotective or overprotective of water quality objectives and designated uses, or when the Tier 1 criteria were unacceptable to regulatory agencies or to regulated entities affected by the SSNC. In Tier 2, the objectives and designated uses are reevaluated using new, additional site-specific data to determine if the uses or objectives affected by nutrients are being attained. The development of Tier 2 SSNC is based on models of nutrient-biological response relationships developed from data on the waterbody under evaluation. Tier 2 provides an opportunity to review and revise the selected nutrient and response variables used to establish the site-specific standard. The review can produce a revised list of variables more relevant to the stated water quality objectives or highly correlated with the probability of major nutrient effects on the waterbody. Data for each variable are required, and more extensive data are required than in Tier 1, including data for calculating seasonal and annual trends, means, and variances for each parameter at each site of interest. Sufficient data are also required to establish the spatial and temporal extent and variability of each variable. Several years of monthly, bimonthly, weekly, or daily data are recommended. If sufficient data are not available, appropriate sampling and information gathering plans must be developed to provide the necessary information.

The goals of data analysis in Tier 2 are to 1) determine if significant relationships exist between nutrient concentrations, biological response variables, and designated uses; 2) quantify such relationships and their uncertainties; and/or 3) identify and evaluate the variables that may confound the development of such relationships. These Tier 2 statistical models can be used to determine the potential benefits of nutrient reduction programs resulting from land management efforts or advanced wastewater treatment. Nutrient reductions should have direct impacts on algal levels, oxygen content, and water clarity, and the magnitude of these changes can be estimated using the empirical models. If significant relationships between nutrient concentrations and Chl *a* are not observed, the existence of possible site-specific relationships between other variables (suspended solids, turbidity, flow, velocity, and grazing) and Chl *a* can be explored.

Three outcomes may result from Tier 2 data analyses:

- One, significant relationships between nutrient concentrations and Chl *a* are identified and SSNC are derived based on those relationships.
- Two, significant relationships between nutrient concentrations and Chl *a* are not established; however, the variables that confound the expected nutrient-Chl *a* relationships are. Significant relationships between these variables and Chl *a* are developed, and SSNC are estimated using these relationships.
- Three, significant relationships between nutrient concentrations and Chl *a* are not identified, nor are confounding variables. In this case, the Tier 2 recommendation is to proceed either to Tier 3 to collect additional data and conduct additional analyses or use the Tier 1 SSNC.

A Tier 3 analysis can be implemented if the Tier 2 process does not provide SSNC that achieve the desired water quality objectives. Tier 3 may also be undertaken if a more comprehensive evaluation of an existing nutrient standard is warranted. Tier 3 methods and approaches can also be used to revise SSNC that are demonstrably inadequate based on the Tier 2 monitoring and evaluation programs. The objectives of the Tier 3 assessment are to

1) produce SSNC, if such criteria were not produced by either Tier 1 or Tier 2 analyses;
2) modify or refine criteria produced by the Tier 2 assessment; or
3) conclude that, given available resources and current scientific understanding of aquatic ecosystems, the development of scientifically credible nutrient criteria that will achieve desired water quality objectives is not feasible.

The Tier 3 assessment process parallels Tier 2 with two important modifications: Tier 3 1) emphasizes the use of site-specific data to support the derivation of SSNC and 2) focuses on process-level evaluation of the site-specific relationships between nutrients and the values of the selected response variables. The Tier 3 assessment continues the Tier 2 emphasis on constructing a mechanistic, process-level understanding among nutrient loadings, nutrients concentrations, and the corresponding magnitude of response by the ecological resources relevant to the stated water quality objectives. Tier 3 may extend the statistical models implemented as part of Tier 2 by incorporating mathematical and computer simulation models developed to characterize the relationships among nutrient loading, subsequent concentrations of nutrients, and the production dynamics of critical populations of aquatic organisms. Thus, Tier 3 offers the possibility of producing SSNC based on state-of-the-art scientific understanding backed by in-depth analysis of site-specific data and limited only by the time and resources available for the assessment.

Tier 3 analyses may provide the following outcomes:

- One, the selected model is successfully implemented for the system of interest. Model-to-data comparisons are of acceptable accuracy and precision from the perspective of the stakeholders. The SSNC are developed using the mechanistic understanding provided by the model.
- Two, the model is not satisfactorily applied because model calculations are not sufficiently accurate to justify using the results or the model to derive the site-specific criteria. However, detailed sensitivity analysis of the model performance may identify key model parameters for which additional data can be required and collected. The

revised model performance then improves to the point of stakeholder acceptability, and the site-specific criteria are developed using the revised model.

- Three, model performance is not improved, despite the collection of additional data. Under these circumstances, the previously developed Tier 1 or Tier 2 site-specific nutrient criteria would likely be retained.

Example applications of the proposed tiered approach for SSNC were constructed for selected streams, rivers, lakes, reservoirs and coastal estuarine systems. Emphasis was placed on freshwater and estuarine surface waters located within the Neuse River, North Carolina watershed. Additional examples of SSNC derivation were based on data obtained for Hinkson Creek, Missouri.

CHAPTER 1.0

INTRODUCTION

1.1 Background and Purpose of Document

The U.S. Environmental Protection Agency has developed guidance for deriving ecoregional nutrient criteria applicable to lakes and reservoirs (U.S. EPA 2000a), rivers and streams (U.S. EPA 2000b), and estuaries (U.S. EPA 2001). For some waterbodies, however, site-specific nutrient criteria (SSNC) rather than ecoregional nutrient criteria may be appropriate. Site-specific criteria apply only to a particular waterbody or to a defined segment of a waterbody. By 2004, if states and Native American tribes have not either adopted the U.S. EPA's ecoregional nutrient criteria and standards or have prepared a plan to develop nutrient criteria (Grubbs, G., Director, Office of Science and Technology, Washington, D.C., memorandum, November 14, 2001), the U.S. EPA may promulgate criteria for them based on the ecoregional analyses and recommendations. Once ecoregional nutrient criteria are published by the U.S. EPA, states, tribes, or other entities may propose their own site-specific criteria if they believe that ecoregional criteria are not appropriate for a particular waterbody. State- or tribal-defined site-specific criteria must be scientifically defensible and approved by the U.S. EPA and state regulatory agencies.

This document provides detailed methods for deriving scientifically defensible SSNC and standards. The document is intended for use by 1) state and tribal regulatory agencies with the authority to derive criteria and standards; 2) regulated entities affected by the criteria, such as dischargers with National Pollutant Discharge Elimination System (NPDES) permits and those potentially affected by nonpoint source controls; and 3) other interested parties or stakeholders. The methods presented here for developing SSNC are meant to complement the U.S. EPA's guidance for developing ecoregional nutrient criteria. In most cases, the application of these methods assumes that ecoregional nutrient criteria have already been developed or proposed by the U.S. EPA. In fact, implementation of ecoregional nutrient criteria that appear to be inappropriate for a specific waterbody will be a prime motivation for developing SSNC.

1.2 Current U.S. EPA Guidance and Example Case Study for Estimating Ecoregional Nutrient Criteria

The U.S. EPA has published three guidance documents for developing nutrient criteria: *Nutrient Criteria Technical Guidance Manual: Lakes and Reservoirs* (U.S. EPA 2000a); and *Nutrient Criteria Technical Guidance Manual: Rivers and Streams* (U.S. EPA 2000b); and *Nutrient Criteria Technical Guidance Manual: Estuarine and Coastal Marine Waters* (U.S. EPA 2001). These documents provide information on the effects of nutrients on aquatic systems, literature reviews on the science of nutrient assessment, and suggested methods for setting ecoregional nutrient standards.

The U.S. EPA recommends two general approaches for setting ecoregional nutrient criteria. The first approach uses data from "reference" waterbodies located within an ecoregion to establish criteria; the second uses data from both reference and nonreference waterbodies located within an ecoregion. Reference waterbodies, ideally, represent conditions where anthropogenic disturbances and pollution are absent (U.S. EPA 2000a, 2000b, 2001). Because most waters have been influenced by human activity, in reality reference conditions are those that are the least affected (U.S. EPA 2000a). The U.S. EPA (2000a, 2000b, 2001) provides general guidance on selecting and classifying reference waterbodies.

In the reference approach, waterbody data are used to establish a cumulative frequency distribution for each variable of interest, such as total phosphorus (TP), total nitrogen (TN), and chlorophyll *a* (Chl *a)*. The *upper* 75th percentile of the distribution for each variable defines the ecoregional nutrient criterion. In the second approach, nutrient data from both reference and nonreference waterbodies in the ecoregion are used to establish a cumulative frequency distribution for each variable of interest. The *lower* 25th percentile of the distribution for each variable defines the ecoregional nutrient criterion.

The U.S. EPA (2000b) guidelines are illustrated using data from reservoirs in the farmland region of Missouri (Table 1-1). The data were collected from 42 reservoirs located in the glacial plains of northern Missouri, which correspond to the northern one-third of the state and is part of the Temperate Forested Plains and Hills Ecoregion within the U.S. EPA's ecoregional classification system. Soils are formed from loess and glacial till from Pleistocene glaciation, and bedrock is primarily limestone and shale. Land use in the region is principally agriculture (85%) and forest (13%). The reservoirs were built for hydropower, flood control, recreation, and water supply. Their limnology is described by Jones and Knowlton (1993). Water clarity is more closely associated with mineral turbidity than algal biomass, and as a result, relationships among nutrients, algal biomass, and water clarity are less obvious than those in natural lakes that have lower levels of suspended solids. For this reason, water clarity is not considered a good measure of lake trophic state in these reservoirs (Jones and Knowlton 1993) and is not considered in this analysis. Variables that may be potential nutrient criteria for this region are TP, TN, and Chl *a*. The lakes were sampled at least three times during the summer seasons of 1989–1999.

Table 1-1. Concentrations (µg/L) of TP, TN, and Chl *a* in Missouri Reservoirs in the Glacial Plains (n = 42) 1989–1999

	Total Phosphorus	Total Nitrogen	Chlorophyll *a*
Mean	51.5	840	16.7
Median	45.5	840	14.0
Minimum	14	410	2.5
Maximum	183	1450	56.2

Identifying "reference" conditions for manmade reservoirs located within an agricultural region can be difficult because of the large land use changes which have occurred in the reservoirs' watersheds. Before they were settled, the glacial plains were assumed to have had a mixture of forest and grassland; the current row-crop agriculture is the major large-scale disturbance to the landscape. Twelve of the 42 reservoirs in the region have less than 16% of the watershed planted in row crops, and the 12 were selected to represent minimally affected conditions within the region.

Following the two approaches outlined by the U.S. EPA (2000b), the ecoregional reference criteria were established. First, using data from only the 12 reference reservoirs, the reservoirs were ranked from low to high according to their TP, TN, and Chl *a* concentration. The upper 75th percentile values were identified as preliminary ecoregional nutrient criteria (Figures 1-1 through 1-3). Based on these data, the criteria would be 32 :g/L for TP, 640 :g/L for TN, and 9 :g/L for Chl *a* (Table 1-2).

Second, using data from all 42 reservoirs in the region, the reservoirs were ranked from low to high according to their TP, TN, and Chl *a* concentrations. The 25th percentile values were identified as the cutoff points for ecoregional nutrient criteria (Figures 1-4 through 1-6). Using the entire data set, the ecoregional criteria would be 26 :g/L for TP, 625 :g/L for TN, and 9 :g/L for Chl *a*. (Table 1-2). Because the results using both approaches are similar, the U.S. EPA guidelines suggest that either set of criteria can be selected or the values can be averaged.

Table 1-2. Ecoregional Nutrient Criteria for Missouri Reservoirs in Glacial Plains, Using Upper 75th and Lower 25th Percentile Concentrations (Data from Jones and Knowlton 1993)

Variable	Ecoregional Criterion Using Upper 75th Percentile (Reference Reservoirs Only) (μg/L)	Ecoregional Criterion Using Lower 25th Percentile (All Reservoirs) (μg /L)
Total Phosphorus	32	26
Total Nitrogen	640	625
Chlorophyll *a*	9	9

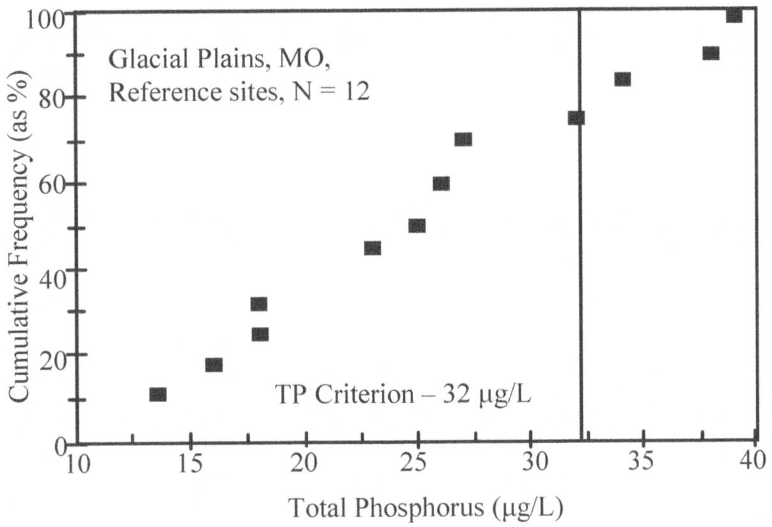

Figure 1-1. Total Phosphorus (μg /L) in Reference Reservoirs in Glacial Plains of Northern Missouri (Vertical Line Indicates 75th Percentile Value) (Jones and Knowlton 1993)

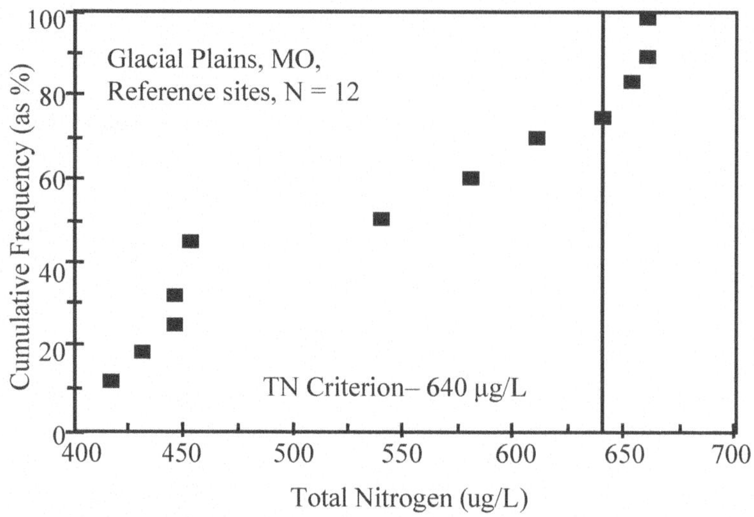

Figure 1-2. Total Nitrogen (µg/L) in Reference Reservoirs in Glacial Plains of Northern Missouri (Vertical Line Indicates 75th Percentile Value) (Jones and Knowlton 1993)

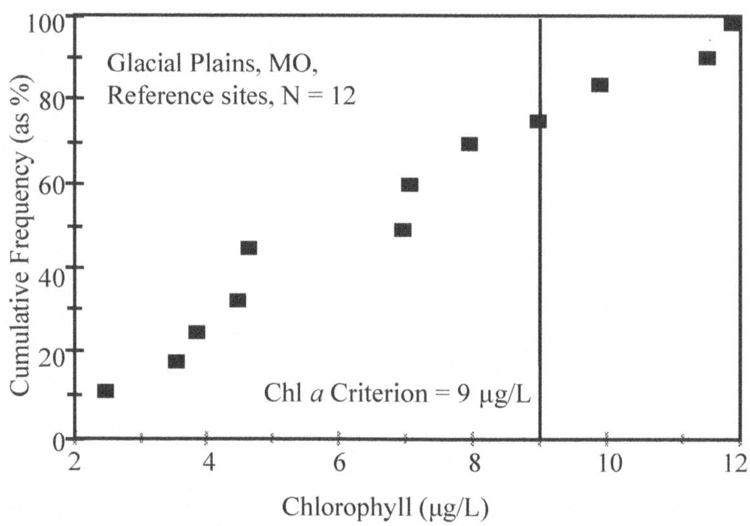

Figure 1-3. Chlorophyll *a* (µg/L) in Reference Reservoirs in Glacial Plains of Northern Missouri (Vertical Line Indicates 75th Percentile Value) (Jones and Knowlton 1993)

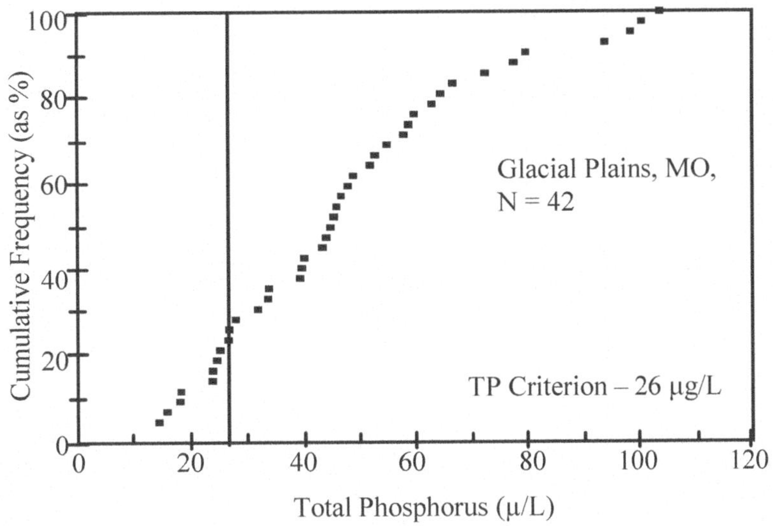

Figure 1-4. Total Phosphorus (µg/L) in 42 Reservoirs in Glacial Plains of Northern Missouri (Vertical Line Indicates 25th Percentile Value) (Jones and Knowlton 1993).

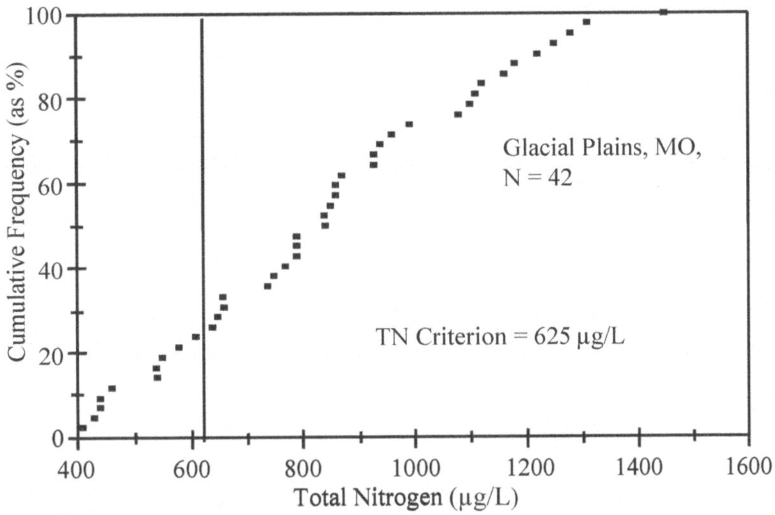

Figure 1-5. Total Nitrogen (µg/L) in 42 Reservoirs in Glacial Plains of Northern Missouri (Vertical Line Indicates 25th Percentile Value) (Jones and Knowlton 1993)

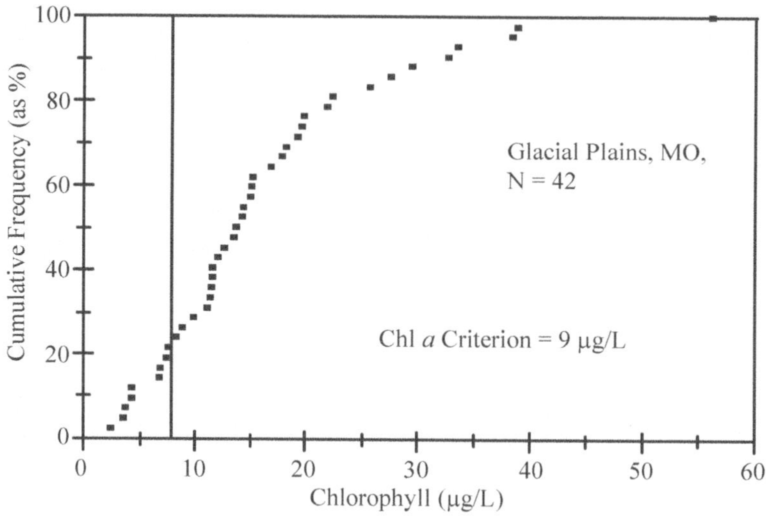

Figure 1-6. Chlorophyll *a* (µg/L) in 42 Reservoirs in Glacial Plains of Northern Missouri (Vertical Line Indicates 25th Percentile Value) (Jones and Knowlton 1993)

1.3 Limitations to U.S. EPA's Approach for Estimating Ecoregional Nutrient Criteria

The U.S. EPA's approach for estimating ecoregional nutrient criteria may be improved or advanced using methods in this document. For example, relationships between nutrients and response variables, such as Chl *a* are not determined in the ecoregional approach. Consequently, in waters where such relationships are weak or may not exist because of confounding factors, management efforts to reduce nutrient concentrations to attain the ecoregional criteria may not produce predicted improvements in Chl *a* concentrations or other response variables. In addition, with the U.S. EPA's approach, relationships between nutrients and designated uses, or other water quality objectives, are not determined. As a result, efforts to reduce nutrient concentrations may not attain the desired designated uses or other nutrient-related water quality objectives.

1.4 A Tiered Approach for Setting Site-specific Nutrient Criteria

1.4.1 Overview

The framework illustrated in Figure 1-7 outlines a three-tiered approach for developing SSNC. This approach is implemented when the potential need for such criteria (see Section 1.4.2) is identified during the decision-making process. The need is determined based on whether 1) the ecoregional nutrient criteria may be underprotective or overprotective of uses or 2) designated uses and water quality objectives have already been attained or are not being attained. The focus of these site-specific methods is to evaluate relationships among nutrients, response variables, and designated uses, and then to use knowledge of these relationships to develop site-specific nutrient criteria.

The tiered approach consists of a sequence of increasingly refined and detailed site-specific methods for deriving nutrient criteria (Sections 1.4.3–1.4.5). The demand for site-specific information also increases with each successive tier. Although the sequence involves increasingly detailed and data-intensive procedures, the process may be initiated at any of the three tiers. The tiers differ in the underlying data and analytical methods used to develop the criteria. The main objective of each tier is to develop site-specific criteria.

Once the SSNC are established, long-term monitoring and evaluation are advisable to ensure that ongoing nutrient management activities are maintaining the desired site-specific water quality and uses (see Section 1.4.6).

The state of science on relationships between nutrients, water quality, and biological responses in rivers and streams, lakes and reservoirs, and estuaries is presented in Chapter 2.0. Chapters 3.0 through 5.0 discuss Tier 1–3 methods. Each chapter addresses each major category shown in Figure 1-7 (the columns of the flow diagram). Chapter 6.0 includes case studies for rivers and streams, lakes and reservoirs, and estuaries.

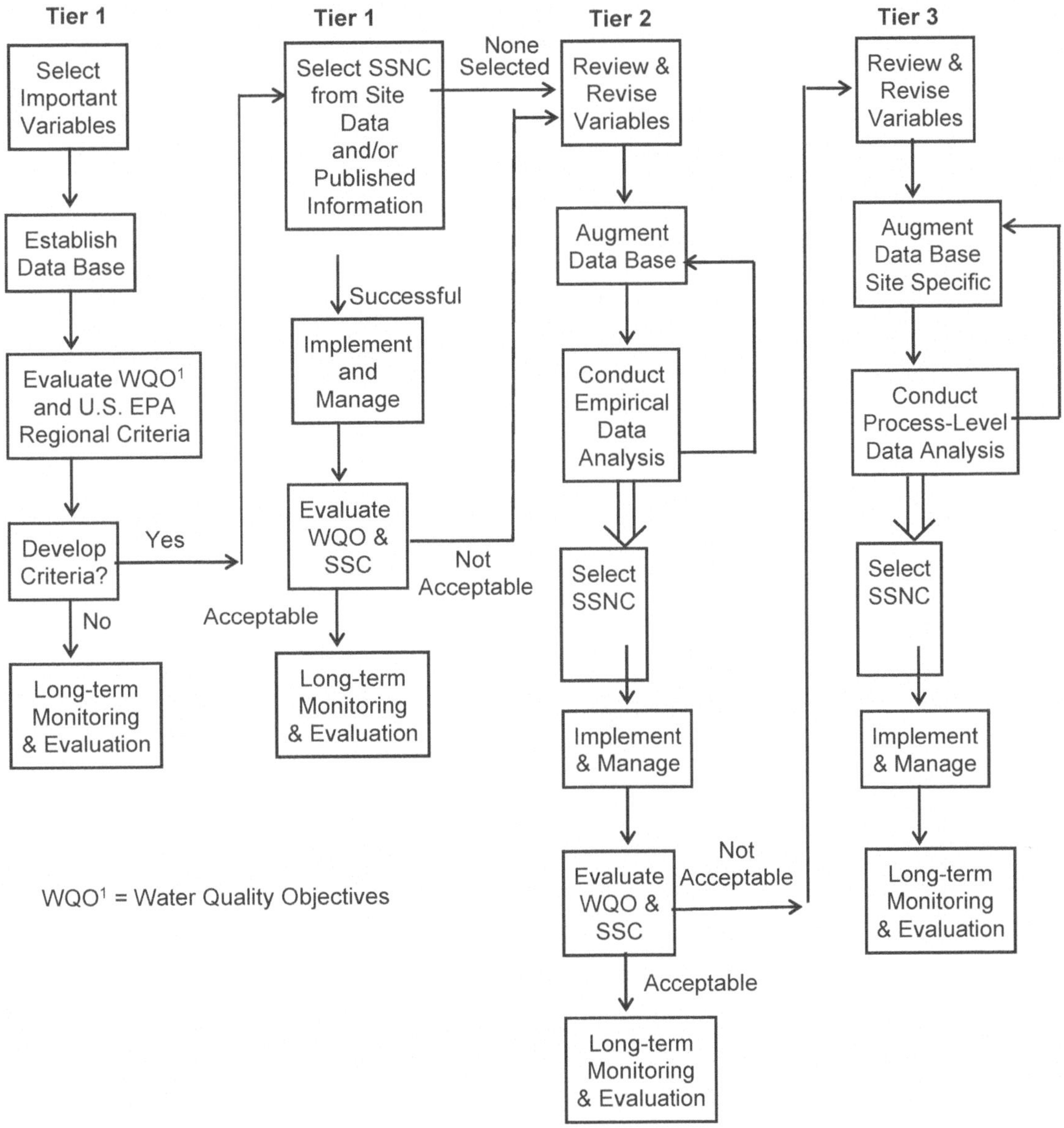

Figure 1-7. Tiered Approach for Setting Site-specific Nutrient Criteria

1.4.2 Problem Description

Since water quality standards are developed and implemented to protect the designated uses of surface waters, the SSNC process must attempt to determine relationships between nutrients and uses of the waterbody. In addition, development of SSNC must consider nutrient issues throughout the watershed, particularly the impact of upstream nutrient concentrations on downstream systems, and not focus strictly on individual waterbodies or segments of waterbodies. For example, SSNC should only be developed and implemented if they will not adversely affect water quality and designated uses in downstream segments and waterbodies.

Similarly, successful implementation of SSNC in a particular waterbody may be hindered or precluded by water quality, habitat degradation, or other factors present upstream or in tributaries to the waterbody that may affect downstream waters. In the beginning of the SSNC process, therefore, the nutrient issues to be addressed should be thoroughly described, especially how they relate to uses of the waterbody and to upstream and downstream waterbodies and their uses. One approach is to begin with the downstream sites and work upstream to establish goals. This approach was successfully carried out by the Chesapeake Bay Program and may be adopted in other areas, like the assessment of hypoxia in the Gulf of Mexico. Initially, this can be done by formulating a series of questions and answers. Examples include the following:

- Is there a nutrient-related water quality problem in the waterbody?
- Do U.S. EPA ecoregional nutrient criteria appear to be overly stringent or insufficient to protect designated or other uses?
- Are uses of the waterbody adversely affected by nutrients?
- What are the uses apparently affected by nutrients?
- How are the nutrient problems manifested?
    - algal blooms?
    - water clarity?
    - dense growths of submerged aquatic vegetation (SAV)?
    - loss of SAV?
    - fish kills?
    - poor fishing?
    - poor tasting drinking water?
- What nutrients are causing/associated with the water quality problems?
    - nitrogen?
    - phosphorus?
    - both?
    - others?
- What are the apparent sources of these nutrients?
    - external, point and nonpoint sources?
    - internal nutrient cycling?
- How is the suitability of nutrient criteria determined?
    - concentration?
    - load?
    - designated uses?
    - U.S. EPA ecoregional nutrient criteria?

- SSNC?
- Do upstream and/or downstream stream segments or waterbodies have similar problems?
- Will upstream or downstream waterbodies or segments be adversely affected by SSNC?
- Are there other, non-nutrient issues that may be contributing to these problems?
  - overfishing?
  - invasive species?
  - poor habitat quality?
  - sedimentation?
  - mineral turbidity?
  - physical or hydrological factors?
  - toxicity?
- Will stakeholders support SSNC?
- Are resources available to support the studies required to develop defensible SSNC?
- Is the problem tractable?

1.4.3 Site-specific Nutrient Criteria Development: The Decision-making Process

All surface waters in the United States have one or more designated uses, many of which are directly or indirectly related to nutrients, and each designated use has associated numerical or narrative water quality standards designed to protect those uses. Figure 1-8 shows how nutrient criteria interface with regulatory policies and actions. SSNC are relevant when ambient water quality monitoring data or other data or observations suggest that ecoregional nutrient criteria are more stringent or less stringent than necessary to protect those uses.

The decision to develop site-specific nutrient criteria is made in relation to existing information concerning nutrient concentrations, ecoregional nutrient criteria, designated uses, and other water quality objectives for the waterbody. Table 1-3 presents the possible results of the associated decision process:

- If one or more designated uses or other water quality objectives affected by nutrients are not being attained and ecoregional nutrient criteria are being exceeded, then nutrients may be adversely affecting local water quality. Such a finding may lead to the waterbody being placed on the 303 (d) list of impaired waterbodies (Figure 1-8). Then a total maximum daily load (TMDL) would be developed to determine load allocations for nonpoint sources and waste load allocations for point sources. Establishing a nutrient reduction target would require an evaluation of relationships among nutrient loadings, subsequent concentrations, and the values of the response variables associated with the desired attainable uses. Reductions in nutrient loads should allow those uses to be attained.

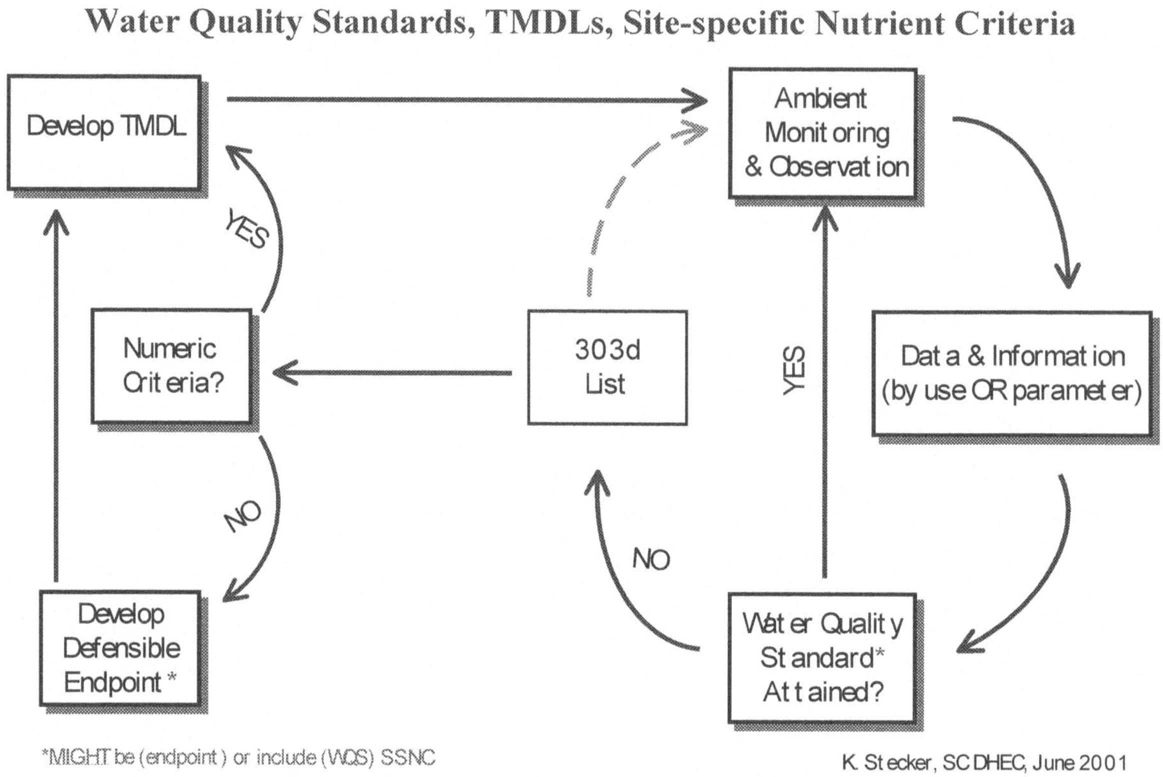

Figure 1-8. How Nutrient Criteria Interface with Regulatory Policy and Actions

Table 1-3. Decision Matrix

Condition	Designated Uses and Water quality Objectives	
	Attained	**Not Attained**
Ecoregional Criteria Exceeded	Criteria are more stringent than necessary Less stringent criteria may be justifiable **Proceed with Tiered Assessment**	Criteria may be adequately protective Adverse impacts present Nutrient loading should be reduced **Develop Nutrient Management Plan**
Ecoregional Criteria Not Exceeded	Criteria appear adequately protective No adverse effects present No changes needed **Implement Monitoring Program**	Criteria are not adequately protective More stringent criteria warranted **Proceed with Tiered Assessment**

- If designated uses or other water quality objectives are not being attained and ecoregional nutrient criteria are not exceeded, then the ecoregional nutrient criteria may not adequately protect the desired uses or objectives. Thus, more stringent SSNC should be developed to protect those uses.
- If designated uses or water quality objectives are being attained and ecoregional criteria are exceeded, then ecoregional criteria are more stringent than necessary. As a result, less stringent site-specific criteria can be developed. *Note that less*

stringent SSNC should be developed only for cases in which such less stringent criteria will not adversely impact downstream waters.

- If designated uses and water quality objectives are attained and ecoregional criteria are not exceeded, then ecoregional criteria are consistent with the designated uses and objectives, and no further action may be necessary. Long-term monitoring should be undertaken to ensure that designated uses and water-quality objectives are maintained.

The following two hypothetical examples demonstrate the use of the decision matrix in determining the need for SSNC.

Example One: A local stakeholder group, concerned with the observations of repeated noxious growths of filamentous green algae (*Chladophora*) along the shoreline of a segment of the Green River and poor tasting drinking water derived from the river, petitions the state water quality agency to develop and implement more stringent nutrient standards. Through collation and examination of historical water quality data collected from several sampling stations in this stream segment (STORET legacy data), the agency determines that TP concentrations, averaged over the months of algal growth for the past 10 years, were consistently < 75 :g/L, compared to the ecoregional criterion in the river of 100 :g/L. Concentrations of TN averaged < 400 :g/L over this same period compared to the ecoregional criterion of 500 :g/L. The agency uses this information and the decision matrix to conclude that the TP and TN criteria historically have not been exceeded, but the goal of achieving high-quality drinking water is not being met, as demonstrated by undesirable odor and taste. According to the decision matrix, although the criteria have not been exceeded, a designated use has not been attained; thus, the development of SSNC may be warranted. As a result, the agency implements the tiered approach to develop potentially more stringent SSNC for TP and TN.

Example Two: A municipal wastewater treatment facility located on a river is faced with upgrading the treatment plant to meet ecoregional nutrient water quality standards. Concentrations of TN in the river averaged 500 :g/L for the past five years, while TP averaged 40 :g/L during this same period. Both concentrations exceed the ecoregional criteria. Further investigation reveals that no incidents of excessive algal growth, fish kills, or problems associated with water clarity, odor, or taste have been reported in the vicinity of the treatment facility or downstream. Complying with the ecoregional water quality criteria would require a significant technical upgrade to the wastewater treatment facility at considerable cost to ratepayers. The municipality uses the available information and the decision matrix to conclude that, even though nutrient criteria have been exceeded, all designated uses and water quality objectives for the river segment have been attained. The municipality decides to use the tiered SSNC process to examine the efficacy of developing less stringent criteria.

1.4.4 Tier 1 Assessment

Tier 1 uses available data or new data, if existing data are inadequate, to develop SSNC. Data from similar waterbodies in the ecoregion may be used as a frame of reference to help in criteria selection. In addition, expert opinion, the technical literature, water quality and other reports, and other ecoregional information may be used to help specify the criteria (Figure 1-7).

Tier 1 methods for developing SSNC start with U.S. EPA's methods for developing ecoregional criteria, with the exception that SSNC development focuses on data from the waterbody of concern, and nutrient and response variable relationships are evaluated. First, cumulative distributions of nutrient data for the waterbody are used to identify percentiles such as the lower 25th percentiles of TN, TP, and Chl *a*, which are used as the proposed SSNC. Then, relationships between nutrients and response variables are evaluated to determine the likely success of the proposed criteria for attaining designated uses and other water quality objectives.

If the available information supports the development of SSNC, Tier 1 advances these criteria for subsequent implementation. Nutrient management plans are designed and implemented to evaluate the performance of the SSNC with regard to achieving the water quality objectives originally outlined by the stakeholders. If the near-term performance evaluation of the SSNC is consistent with achieving and maintaining designated uses and other the water quality objectives, the Tier 1 assessment concludes with the development of a longer term monitoring plan and continued evaluation of the SSNC performance.

If the performance of the specified Tier 1 SSNC relative to the designated uses and other water quality objectives proves unsatisfactory, the assessment should proceed to Tier 2. Under certain circumstances, particularly when high quality data and models are available, Tier 2 may be bypassed, and the assessment can proceed directly to Tier 3. If adequate data are available, Tier 1 SSNC can be developed relatively quickly. If extant data are inadequate and new data collections are required, however, the development of Tier 1 SSNC may require several years. Tier 1 is discussed in detail in Chapter 3.0.

1.4.5 Tier 2 Assessment

The Tier 2 assessment expands on the Tier 1 approach and consists of the following activities:

1. review and possible revision of the selected nutrient and response variables,
2. augmentation of the Tier 1 information base,
3. empirical analysis of possible quantitative relationships between nutrients and the response variables,
4. development of Tier 2 SSNC,
5. implementation of the criteria and the associated nutrient management plan,
6. evaluation of Tier 2 SSNC performance, and
7. long-term monitoring of SSNC to prove that the desired water quality objectives are being met.

Tier 2 is discussed in detail in Chapter 4.0.

1.4.6 Tier 3 Assessment

The Tier 3 process parallels Tier 2 with two important modifications (Figure 1-7). First, Tier 3 emphasizes the collection of site-specific data necessary to support the derivation of SSNC. Second, Tier 3 analyses require a process-level evaluation of the site-specific relationships between nutrients and the selected response variables. Correspondingly, Tier 3 offers the possibility of producing SSNC based on state-of-the-art scientific understanding supported by in-depth analysis of site-specific data limited only by the time and resources committed to the assessment. Not surprisingly, Tier 3 is also the most costly and time consuming of the three tiers. Depending on the types of data collected, Tier 3 SSNC may require several years to a decade to develop. Tier 3 is discussed in detail in Chapter 5.0.

1.4.7 Long-term Monitoring

Once SSNC have been established, long-term monitoring to evaluate their effectiveness should be established. Long-term monitoring will also help determine if designated uses have been attained or if other water quality objectives have been achieved and alternative SSNC are unnecessary. In both cases, long-term monitoring should continue to determine if nutrient management activities are maintaining the desired site-specific water quality objectives and designated uses.

Even when SSNC are not developed, monitoring should continue until either 1) water quality agencies change the local water quality objectives or desired uses or both, or 2) the monitoring data demonstrate that previously specified water quality objectives are no longer being met. Both circumstances may support the need to develop SSNC, including reevaluation of the decision matrix and possible implementation of the tiered assessment approach. One result of this process may be the design and implementation of a new or revised monitoring program.

1.4.8 Role of Stakeholders in the Site-specific Nutrient Criteria Process

Stakeholders have an important role in the development of SSNC. Although the development of water quality criteria and standards is a scientific process, stakeholders may provide input about water quality concerns related to nutrients, comment on proposed criteria and standards, and contribute ideas on implementation. Appendix A includes guidelines for stakeholder involvement in the development and implementation of SSNC.

CHAPTER 2.0

STATE OF THE SCIENCE

2.1 Introduction

Excessive nutrient loadings have degraded the quality of surface waters in the United States. The U.S. EPA (1996) reported that excess nutrients contributed to the impairment of water quality in 40% of the nation's assessed rivers and streams. Until recently, quantitative, numerical standards for nutrients remained largely undeveloped. Historically, efforts to improve water quality in surface waters have focused on regulation of point sources, such as wastewater discharges and toxic chemicals; nonpoint sources of nutrients were not regulated. In the few situations where nutrient discharges were regulated to meet numeric criteria, the primary objectives were to reduce potential toxic effects on aquatic organisms (effects of ammonia) and humans (effects of nitrate). Nutrient enrichment, or eutrophication, was not the central issue around which numeric criteria were developed.

The development of SSNC for surface waters should be based on a quantitative understanding of the physical, chemical, and biological processes that determine the ecological responses of aquatic systems to excessive nutrients. Such understanding should address nutrient point and nonpoint source loading; the physical and hydrodynamic processes that determine nutrient transport, distribution, and fate within aquatic systems; the complex mechanisms that govern the chemical form of nutrients within the water column and sediments; and the biological and ecological transformation of nutrients in the production of new biomass (National Academy of Science 1969). SSNC require a quantitative relationship between concentrations and loads of nutrients for specific designated uses.

This chapter highlights particular aspects of the eutrophication process that appear important in deriving SSNC. A thorough and comprehensive review of the extensive eutrophication literature, however, is beyond the scope of this report. For additional information on the effects of nutrient enrichment see Bricker et al. (1999). The following sections discuss current understanding of nutrient-related issues for rivers, streams, lakes, reservoirs, and estuaries. Potentially useful tools for developing SSNC are also described.

2.1.1 Common Nutrient Measurements

The development of SSNC requires accurate and precise measurements of nutrients in aquatic systems. Nitrogen (N) and phosphorus (P), the primary nutrients of concern, occur in inorganic, organic, particulate and dissolved forms. The most common measurements of N and P include TN, TP, total inorganic nitrogen (TIN), dissolved inorganic nitrogen (DIN), and soluble reactive phosphorus (SRP).

- TN includes all inorganic, organic, dissolved, and particulate forms of nitrogen; TIN includes all inorganic forms of nitrogen (NO_2, NO_3, and NH_4); DIN includes all dissolved inorganic forms of nitrogen.

- TP includes all inorganic, organic, dissolved, and particulate forms of phosphorus; SRP is the best measurement of dissolved orthophosphate.

Plants preferentially use dissolved inorganic forms of nutrients. Therefore, measurements of dissolved inorganic nutrients sometimes are preferred for establishing relationships between nutrient concentrations and aquatic plant growth, especially in natural lakes. In some lakes, particularly those with short residence times, the predictive relationships may be strong. In many waterbodies, however, measurements of TN and TP are often more highly correlated with measurements of algal biomass (Chl *a*) than are TIN or SRP.

Additional nutrients of concern, particularly in estuaries, include potassium, carbonate, calcium, magnesium, sodium, silica, and sulfate. For example, measures of sulfate reduction in anaerobic respiration in estuarine sediments suggest that 70% of the energy of net primary productivity may pass through reduced inorganic sulfur compounds (Howarth et al. 1983). Important trace elements include iron, manganese, molybdenum, cobalt, and zinc. Certain organic compounds are also important, including vitamins and simple sugars.

2.1.2 Limiting Nutrient Concept

The nutrient demanded by growing plants that exceeds availability to the greatest extent can be defined as the nutrient that limits growth, or the "limiting nutrient." Plants vary in their relative ability to accumulate nutrients from their surrounding environment (water column or sediments); plants can also accumulate internal stores of nutrients that exceed their immediate needs for growth (luxury consumption) to provide for future growth and reproduction. In addition, environmental factors other than nutrients can constrain plant growth, for example, light intensity, temperature, and grazing pressure. Thus, the limiting nutrient concept is biologically complex, modified by many external factors, and dynamic in space and time.

Both N and P can be limiting nutrients for plant growth in rivers and streams; plant growth in lakes is frequently limited by available P, while estuarine primary productivity is often N limited. The limiting nutrient can change seasonally. For example, the availability of dissolved silica (Si) can limit production during spring and early summer when diatoms commonly dominate the flora of many aquatic systems. Dissolved inorganic P may be limiting during the earlier months of plant growth, with a subsequent shift to N limitation. Atomic ratios of N:P higher than 16:1 suggest P limitation, while ratios lower than 10 often indicate N limitation. Alternatively, a micronutrient or an essential vitamin can become the limiting nutrient.

The important point is that the development of SSNC may require a detailed quantitative characterization of spatial-temporal variations in the identity of the limiting nutrient and its limiting concentration. Criteria may differentially emphasize temporal loading rates and subsequent concentrations of P, N, and Si, as well as availability of important micronutrients. Realistically, criteria will likely focus on macronutrients. The dynamics of plant growth limitation by micronutrients and other factors (light, temperature, grazing pressure), however, may influence the perceived effectiveness of criteria that are implemented for N and P.

2.1.3 Ecological Responses to Nutrient Enrichment

Eutrophication can cause a variety of adverse human health effects along with undesired ecological effects in aquatic ecosystems (U.S. EPA 1998). The values for SSNC will derive mainly from the associated biological and ecological responses to excessive nutrient availability. Under certain circumstances, however, site-specific criteria may actually allow for increases in nutrient loading (to increase local production of fish or other desired aquatic organisms). The following sections describe different ecological responses to enrichment that may be used to develop SSNC.

2.1.3.1 Aquatic Plant Biomass

Water quality standards for nutrients may be developed to control the undesired growth of algae or other aquatic plants. Phytoplankton biomass is routinely measured as Chl *a*, in vitro fluorescence, or ash-free dry mass; it is also estimated from direct cell counts. Biomass of phytoplankton is routinely reported in volumetric units (Chl *a* as mg/L), while the biomass of attached algae (periphyton) is given on an areal basis (mg/m^2 Chl *a*). Rates of algal production can be measured using methods that include radioisotopes (^{14}C) or measurements of dissolved oxygen (DO). Such measures can be used to define acceptable values of algal biomass or productivity in the specification of SSNC.

Nutrient enrichment can also shift the algal community composition to less desirable species (noxious blue-greens or to *Cladophora*). These species may further directly impair water quality or result in subsequent shifts in the community composition of ecological guilds (zooplankton) that graze on phytoplankton or periphyton.

The potential nuisance growth of submerged aquatic vegetation (SAV) may be used to establish SSNC. The abundance of SAV is commonly measured as numbers of stems or biomass (g dry mass, ash-free dry mass) per m^2. In the presence of excessive nutrient loading, diverse assemblages of SAV may be replaced by dense monotypic plant beds (*Myriophyllum*, *Ceratophyllum*). Thus, the diversity of plant species characteristic of local SAV communities may also form the basis for developing SSNC. In shallow lakes and estuaries, other effects may occur. Excess nitrogen allows phytoplankton and epiphytes to out compete SAV for light, leading to declines in SAV. Loss of SAV leads to loss of valuable habitat for many species.

2.1.3.2 Water Quality Parameters

Measurements of certain water quality variables can be used to develop and monitor the effectiveness of SSNC. Desirable ranges or unacceptable values of several interrelated measures of water clarity (Secchi depth), turbidity, suspended particulate matter, and color can be used to specify criteria. Their use in defining criteria requires establishing site-specific relationships between nutrient loading and these measures of water quality. Conversely, measurements of nonvolatile suspended solids (turbidity caused by suspended minerals) can be used to determine the magnitude of reduction in photosynthesis caused by shading.

The excessive production of algae and other aquatic plants can increase pH and initially increase DO concentration. However, highly productive systems also lead to increases in plant and community respiration, thus reducing DO concentrations, particularly during the night or early morning, to levels that are stressful or even lethal to aquatic organisms. Decreases in pH also may occur when respiration exceeds production. High rates of plant respiration or the decomposition of large amounts of organic matter (algal biomass) can cause the release of ammonia, methane, or hydrogen sulfide from sediments that become anoxic. These releases may contribute to additional mortality in fish or other aquatic organisms.

Other less quantitative measures of nutrient loading and subsequent excessive or nuisance plant growth include odor and taste of drinking water, as well as the obvious formation of surface algal films or the odor associated with the decomposition of dense algal blooms.

2.1.3.3 Fish Production and Community Structure

The increases in primary production stimulated by nutrient enrichment can translate to increased production and biomass of other aquatic food web components. For example, increases in the population sizes of fish are a common indirect response of aquatic systems to nutrient enrichment. Depending on the particular growth requirements and trophic interactions, individual species of fish may respond differentially to increases in primary production and associated changes in water quality. Increases in algal abundance result in lower concentrations of DO can increase overall fish abundance but shift the relative biomass towards species such as carp that can use the increase in algae and tolerate the lower DO environment. The response of the fish community in the central and eastern basins of Lake Erie to cultural eutrophication is an example of this phenomenon: native salmonids and forage species were replaced by less desirable species, although the estimated overall biomass of fish apparently increased.

At lower to intermediate degrees of nutrient enrichment, the existing fish populations may simply increase in size without appreciable changes in community structure. This response may reasonably be expected for communities dominated by phytophageous fish species that are the main prey of the local piscivorous fish species in comparatively short aquatic food chains. This overall increase in fish biomass may be facilitated if the forage species are relatively insensitive to shifts in algal species composition (increased production of blue-green algae) that result from nutrient enrichment.

Finally, local public interests may want to increase the biomass of commercially or recreationally valuable fish species. This water quality objective may require an increase in the permissible concentrations of nutrients necessary to directly or indirectly fuel the desired increases in fish production. In light of these possible responses to eutrophication, SSNC may be derived from stakeholder water quality objectives concerning the abundance and species composition of fish.

2.1.3.4 Aquatic Invertebrate Production and Community Structure

Many of the same issues outlined above for the responses of fish to nutrient enrichment apply to the production and community structure of aquatic invertebrates. These responses may be used to develop SSNC based on water quality objectives concerning aquatic invertebrates.

Low DO concentrations that result indirectly from nutrient enrichment can reduce species diversity in communities of benthic invertebrates. Community composition can shift toward species, such as tubificids or chironomids, which are more tolerant of low DO conditions. These species may be less desirable from an ecological or social perspective. Such shifts in benthic community structure may may cause subsequent changes in fish community structure, particularly for those fish that specialize in feeding on selected invertebrate prey that have been displaced by prey species more tolerant of eutrophic, anaerobic conditions.

Shifts in invertebrate community structure can also result from nutrient-related shifts in the relative abundance of the algal species that constitute the assemblages of phytoplankton or periphyton—a primary food source for invertebrate grazers.

2.1.3.5 The Microbial Loop

Under certain environmental circumstances, significant amounts of carbon (or energy) flow through those bacteria, protozoa, micro-algae, and other unicellular flora and fauna that constitute the "microbial loop" in aquatic ecosystems (Stockner and Porter 1989). More traditional discussions and studies of the impacts of nutrient enrichment have emphasized changes in the distribution and abundance of larger phytoplankton, zooplankton, benthic invertebrates, and fish. Recent investigations highlight the need to consider nutrient-induced changes in the relative importance of the microbial loop in characterizing the impacts of cultural eutrophication.

Given a limited quantitative understanding of the ecology of the microbial loop, impacts on this potentially important ecosystem component are unlikely to become a primary focus in developing SSNC. However, the potential importance of the loop in influencing the functional response of other system components to alterations in nutrient availability may have to be addressed when developing SSNC.

2.1.3.6 Habitat Quality and Quantity

Excess nutrients may affect the quality and availability of habitat for fish and wildlife. For example, increased sedimentation of particulate organic matter may reduce the quality and quantity of fish-spawning habitat for those species that require gravel or rocky substrate. Decreased water clarity resulting from elevated phytoplankton productivity may discourage foraging by wading birds such as the great blue heron. Shading and competition from phytoplankton can also lead to losses of SAV. Alternatively, dense beds of SAV stimulated by nutrients may reduce riparian mammal (for example, mink or raccoon) access to desired aquatic prey species. Accelerated sedimentation of shallow lakes, backwaters, and impoundments by increased loads of particulate organic matter can reduce the availability of large acreages of habitat for a variety of riparian and aquatic organisms.

More than 100 years of in-depth, quantitative study of the changes in aquatic ecosystem structure and function resulting from elevated nutrient supplies has produced a complex picture of impacts that can be measured across a wide range of ecological scales and complexity (microbial dynamics to large-scale habitat alterations). Selection of environmental response variables that are appropriately scaled and most relevant to stated site-specific water quality objectives will depend on knowledge and understanding of the complex relationships between nutrient availability and the dynamic ecology of aquatic ecosystems.

Many aspects of eutrophication apply to surface waters in general, for example, increased primary production, decreased water clarity, shifts in community structure, and increased decomposition of organic matter and the associated DO depletion. At the same time, scientific issues specific to a variety of surface waters (streams and rivers, lakes and reservoirs, and estuaries) may have to be addressed by 1) selecting appropriate response variables, and 2) developing quantitative relationships between nutrient availability and responses of selected variables, including those that affect designated uses. Both are central to the establishment of SSNC.

2.2 Rivers and Streams

2.2.1 Overview

The derivation of SSNC for rivers and streams poses particular challenges. These lotic systems, dominated by the unidirectional flow of water, are strongly influenced by local and regional hydrology. Streams and rivers are functioning ecological systems that can be directly affected by excessive nutrient inputs. These systems also serve as nutrient conduits to lakes, reservoirs, and estuaries, as well as to other lotic systems. The development of SSNC must address streams and rivers both as ecosystems and as vehicles for downstream nutrient transport.

2.2.2 Significant Issues

2.2.2.1 Nutrient Effects

Excess levels of nutrients in rivers and streams can cause a variety of adverse human health and ecological effects, although, in some cases nutrients may have beneficial effects (U.S. EPA 1998). Excess nutrients cause eutrophication of rivers and streams, the increased primary production and growth of aquatic plants, including algae and macrophytes. In shallow, clear rivers and streams with higher water velocities, benthic periphyton tend to be the most important primary producers, while in some shallow rivers and streams with lower velocities, macrophytes also may be important. In deeper, slower rivers and streams, sestonic algae (phytoplankton) tend to be the most important primary producers. Some of the adverse effects of eutrophication of rivers and streams include the following (U.S. EPA 2000b):

- Blooms of undesirable algal species, including attached algae such as *Cladophora* or floating algae such as the blue-green *Aphanizomenon flos-aquae*. These blooms may cause aesthetic and ecological problems and interfere with recreational activities, such as boating, swimming, and fishing.
- Dense growths or stands of submerged aquatic macrophytes that can also cause aesthetic and ecological problems and interfere with recreational activities.
- Changes in fish, invertebrate, and plant species composition and abundance, often favoring undesirable species.
- Taste and odor problems in drinking water, as well as contributing to the production of carcinogenic trihalomethanes in chlorinated drinking water.
- Diurnal or nocturnal increases in pH that can increase the toxicity of ammonia and some metals to aquatic life.

- ♦ Low DO concentrations, including anoxia, especially during the night or early morning, resulting from increased plant and microbial respiration rates. Anoxic sediments can release ammonia, methane, and hydrogen sulfide. These phenomena can cause or contribute to stress and mortality of fish and other aquatic life, in particular sensitive, demersal, early life stages.

In some cases, increases in nutrient concentrations in rivers or streams may have positive effects on aquatic resources. The most common example is increased productivity of populations of commercially or recreationally important fish species.

2.2.2.2 Limiting Nutrients

Both N and P can be limiting nutrients for plant growth in rivers and streams (U.S. EPA 2000b). Phosphorus most commonly is limiting, except in areas with naturally high levels of phosphorus and in streams receiving agricultural or municipal wastewater discharges containing high levels of phosphorus and low N/P ratios. In those waters, N availability can limit plant growth.

2.2.2.3 Biological Response Variables

Plant biomass in rivers and streams is usually measured as Chl *a* and expressed in units of mg/m^2 for periphyton or :g/L for phytoplankton (U.S. EPA 2000b). Other common measurements include cell counts, species composition and densities, and measurements of primary production and respiration rates. Sometimes, measurements of turbidity or transparency that could be correlated with phytoplankton density are used as indirect measures of phytoplankton density in lakes and reservoirs. In rivers and streams, however, turbidity and transparency are often poor indicators of phytoplankton density, because high turbidity and low transparency can be caused by high concentrations of suspended sediments or dissolved organic carbon.

2.2.2.4 Other Limiting Factors

Factors other than nutrients that may affect primary production in rivers and streams include turbidity, shading, water depth, water velocity, discharge, grazing, and type of benthic substrate (U.S. EPA 2000b). Turbidity and shading reduce photosynthesis by decreasing the quantity of solar radiation penetrating the water. Solar radiation decreases with increasing water depth and reduces both periphyton and phytoplankton photosynthesis. Water velocity, floods, and scour influence periphyton growth and accumulation in shallower rivers and streams; in general, periphyton biomass decreases as water velocity and scour increases. In deeper rivers and streams, water velocity and flow rate influence the rate at which phytoplankton cells are flushed from the system. In general, higher flow rate and stream velocity result in lower phytoplankton density. Drought, low flows, or desiccation limit periphyton growth in streams in some regions by reducing the amount of inhabitable substrate. Grazing by macroinvertebrates can reduce periphyton and phytoplankton biomass. Finally, the type of benthic substrate influences the periphyton biomass. Unstable substrates such as sand, silt, and muck tend to have lower periphyton biomass than stable substrates such as rubble or cobble.

In rivers and streams where one or more of these factors limit primary production, nutrient concentrations may not be highly correlated with periphyton or phytoplankton biomass and, in some cases, no significant correlation exist. Consequently, changes in nutrient concentrations may not always lead to changes in algal biomass (U.S. EPA 2000b). These factors are important considerations in the development of SSNC and in the management of waters with elevated nutrient concentrations.

2.2.2.5 Stream Order and the River Continuum

Stream order classifies a channel with reference to its location in a network of tributaries that ultimately defines a river system. Order ranges from the smaller, first-order streams that drain small sub-catchments through higher (fourth and fifth) order larger tributaries that connect to the main river channel. The physical dimensions of a channel and its spatial relation to other channels and tributaries are important in determining the hydraulic and ecological characteristics as well as the watershed area of a stream or river.

The river continuum theory (Vannote et al. 1980) describes a pattern of structural and functional changes that occur during the progression from low- to high-order streams and tributaries to the main river channel. The river continuum theory posits, in part, that

- current velocities (m/s) decrease with increasing stream order;
- flows (m^3/s) increase with increasing stream order;
- lower order streams are primarily lotic in their ecological characteristics;
- lower order streams are predominantly heterotrophic, dependent on allochthonous inputs of organic matter;
- higher order streams and rivers are increasingly lacustrine in their ecological characteristics; and
- higher order streams are increasingly autotrophic with significant autochthonous production of organic matter.

The river continuum theory has also been applied to describe reservoirs. The upper reaches of reservoirs are more riverine in nature, with narrower channels, higher current velocities, and greater dependence on external inputs of organic matter. The lower sections proximate to the dam are typically wider and deeper, with slower current velocities. These lake-like regions of the reservoir are often more autotrophic. These two portions of a reservoir are separated by a transition segment of variable length in which the lotic characteristics of the upper reach become increasingly lentic.

Experience indicates that different river systems appear more or less consistent with these projected patterns, but not all streams and rivers follow the continuum and other theories have been offered to describe lotic systems (the serial discontinuity concept, Ward and Stanford 1983). Importantly, the development of SSNC may require an evaluation of the system of interest in relation to the tenets of the continuum or other lotic system theory. In addition, human influence can play an important role in stream structure, particularly in higher order streams.

2.2.2.6 Discharge and Rating Curves

The temporal patterns of water flow define the discharge regime for rivers and streams. Discharge can be important in the derivation of nutrient criteria. Flow regimes are recognized in established water quality standards that are referenced, for example, to the $7Q_{10}$, which is the seven-day low flow with a recurrence interval of once in ten years. SSNC should be developed in the context of a relevant period of record that defines historical high, low, and normal range of flows for the stream or river of interest.

The rating curve defines a stream or river hydraulic response to precipitation, the continuous rise and fall of discharge as floodwaters move past a fixed point in the river. This curve is important in understanding the watershed loading and subsequent transport of nutrients for rain events of different magnitudes and duration. Quantitative knowledge of the response of streams and rivers to patterns of precipitation may be important in 1) deriving reasonable criteria for nutrients, 2) evaluating the efficacy of nutrient management plans, and 3) designing monitoring programs.

2.2.2.7 Nutrient Spiraling

The dominant advective nature of lotic systems has important implications for describing the internal cycling of nutrients in rivers and streams. The classical model of nutrient cycling in lakes describes the uptake of dissolved nutrients by planktonic algae, the eventual sedimentation of these nutrients in the form of particulate organic matter, and the remineralization of nutrients back into the dissolved inorganic nutrient pool as the organic matter decomposes. Adding the advective term to this cycling model for rivers and streams produces a nutrient "spiral" in which remineralized nutrients are translocated downstream before being taken up again by plants (Newbold et al. 1981). The "spiraling length" measures the distance of nutrient translocation. Spiraling length is determined by the temporal scales and magnitudes of biological and chemical processes involved in nutrient uptake and remineralization, as well as the hydrodynamic characteristics such as discharge or current velocity of the stream or river. Clearly, spiraling lengths are not constants but vary in relation to allochthonous nutrient inputs, system productivity, and flow patterns.

Spiraling length can be important in anticipating the implications of upstream nutrient management on the likely effectiveness of SSNC. Spiraling length is also important in projecting the downstream impacts of such criteria.

2.2.2.8 Extreme Events

Compared to typical patterns of circulation in lakes and controlled flows through reservoirs, streams and rivers can vary significantly compared to their average patterns of discharge. Floods can transport large amounts of nutrient-laden sediments and large particulate matter, with an accompanying dislocation (scouring) of periphyton, while desiccation from droughts or diversions of stream flow can decimate resident communities. These extreme and uncontrollable events may have to be factored into the development of SSNC, particularly in considering the efficacy of established criteria following such an event. The occurrence of extreme flow events may also force the redesign of long-term monitoring programs used to evaluate the effectiveness of SSNC.

2.2.2.9 Dams

The future damming of streams and rivers can force the reevaluation of previously established SSNC. Alterations in annual discharge regimes of rivers downstream from new dams may invalidate the assumptions of flow patterns needed to sustain desired water quality objectives in relation to present or anticipated nutrient loading. Substantial alterations in downstream water quality, both positive and negative, may result from damming, depending on whether impounded surface or bottom water is released. Ambient nutrient concentrations and other factors that govern biological productivity and response to nutrients such as water temperature or DO can change dramatically upon damming a stream or river. Of course, the removal of dams can also have a great influence on the stability and structure of streams.

2.3 Lakes and Reservoirs

The diversity in size, shape, and geological origin of lakes, among other factors, contributes to the difficulties in establishing meaningful SSNC for lakes and reservoirs. Reservoirs combine many of the structural and functional features of lotic and lentic systems, which adds to the challenges in developing SSNC for these manmade systems.

2.3.1 Overview

During the past 100 years, the science of limnology has developed a trophic state system to characterize lakes and relate many biological, chemical, and physical features of lakes and reservoirs to their trophic state. Early on, the science was descriptive and separated lakes into three major categories (oligo-, meso-, and eutrophic) that define a fertility continuum from least productive to most productive lakes. This effort originated in Europe, where regional lake studies showed that phytoplankton production, and the biology of the lake as a whole, was determined by nutrient levels. Lake nutrient concentrations are a function of the geological and nutrient loading characteristics of the lake's watershed, as modified by lake morphometry. Lakes with large inputs of nutrients tend to be *eutrophic* (well nourished and productive), while those with moderate or poor nutrient inputs are *mesotrophic* and *oligotrophic*, respectively. The general characteristics of oligotrophic and eutrophic lakes are outlined in Table 2-1. For many uses related to recreation, aesthetics, and water supply, oligotrophic lakes are preferred, although many naturally eutrophic lakes are highly valued regional resources.

The strong influence of the geographic setting of a lake on its chemical and biological characteristics has long been recognized (Rawson 1939). The primary determinant of lake trophic state is the geological/chemical nature of the edaphic setting, which is further modified by lake basin morphology and hydrology (timing and volume of runoff). Human influences such as agriculture and urbanization contribute to the eutrophication process, causing waterbodies to become more eutrophic through nutrient loading in excess of inherent background levels determined by local edaphic conditions. Changes in ecosystem structure and function resulting from these anthropogenic additions of nutrients to lakes are the impetus for setting nutrient criteria for the purpose of lake protection.

Table 2-1. General Characteristics of Oligotrophic and Eutrophic Lakes (After Rast and Lee 1978)

Characteristic	Oligotrophic	Eutrophic
Plant Production	Low	High
No. of Algal Species	Many	Few
Dominant Algae	Diatoms	Blue-greens
Aquatic Rooted Plants	Sparse	Abundant
Hypolimnion Oxygen	Present	Absent
Dominant Fish	Deep-dwelling, coldwater fish, such as trout, salmon and cisco	Surface-dwelling, warmwater fish such as pike, perch, and bass; also bottom-dwelling fish such as catfish and carp
Water Quality for Recreation, Domestic and Industrial Use	Good	Poor

Nurnburg (1996) reviewed the literature on trophic state classification and showed that lake trophic state is closely tied to nutrients. The concept of trophic state was summarized by Carlson and Simpson (1996). Trophic state classification has been based on surface water quality in the open water of lakes (Carlson 1977, Dillon and Rigler 1975) or has included oxygen content of the deep (hypolimnetic) water (Walker 1979). Since the inception of lake trophic state, additional categories—such as ultraoligotrophic and hypereutrophic—have been added to extend and better characterize the continuum of lakes along a fertility gradient that spans several orders of magnitude in nutrient concentration. Among lakes, phosphorus and nitrogen concentrations are closely correlated (Downing and McCauley 1992), and either N or P can serve as a measure of a lake's trophic state (Table 2-2). Nurnberg (1996) concluded that average epilimnetic phosphorus during summer is a better predictor of trophic state than phosphorus averages for other seasons. Limnologists agree rather closely regarding the limits set to separate lakes into oligotrophic, mesotrophic, eutrophic and hypereutrophic categories (Table 2-2). Variations among nutrient limits suggested by various investigators result from regional differences and the heterogeneous nature of the data. Generally, most lakes with mean depths >10m are oligo- or mesotrophic, while shallower lakes are typical of higher trophic state.

Table 2-2. Trophic State Limits Proposed by Various Authors (Units are µg/L)

Variable	Oligotrophic-mesotrophic	Mesotrophic-eutrophic	Eutrophic-hypereutrophic	Reference
Total P	10	30	100	Nurnberg (1996)
	15	25	100	Forsberg and Ryding (1980)
	10	25	100	Jones and Knowlton (1993)
Total N	350	650	1200	Nurnberg (1996)
	400	600	1500	Forsberg and Ryding (1980)
	300	500	1200	Jones and Knowlton (1993)
Chl *a*	3.5	9	25	Nurnberg (1996)
	3	7	40	Forsberg and Ryding (1980)
	3	7	40	Jones and Knowlton (1993)

Algal biomass, measured as algal chlorophyll, is the most direct indicator of lake trophic state and is directly related to water clarity (Carlson 1977). In lakes with low levels of mineral turbidity, water clarity is a direct function of algal biomass and is readily interpreted by the general public as a simple measure of water quality. An objective of lake management is to regulate nutrients to control levels of algal biomass and improve water clarity.

Summer hypolimnetic oxygen is also used to define the trophic state of lakes because of its importance to fish and the release of nutrients from anoxic lake sediments. Nurnberg (1996) has proposed the anoxic factor (AF) (days of oxygen depletion per summer) or areal hypolimnetic oxygen depletion rate (AHOD) (mg · m^2/day ^{2}) for the classification of hypolimnetic oxygen content. Both AF and AHOD are highly correlated with nutrients. Nurnberg (1996) also points out that morphometry plays a role in hypolimnetic oxygen depletion and that many lakes have anoxic deep waters despite oligotrophic nutrient and biomass levels. In these systems, anoxia is not caused by eutrophication.

The abundance and richness of coldwater fish is a function of lake trophic state. Oligotrophic lakes have as many as six or more coldwater species, mesotrophic lakes have few or none, and eutrophic lakes typically do not support coldwater fish because of anoxia in the hypolimnion (Nurnberg 1996). However, Oglesby (1977) found high correlations between fish yields (harvest) from lakes and summer chlorophyll levels. Yurk and Ney (1989) found strong correlations between phosphorus and fish standing stocks in lakes. As lakes become more eutrophic, the standing crops, productivity, and yields of fish increase. This feature has been identified as one of the positive effects of eutrophication (Bachmann et al., 1996). However, as eutrophication proceeds, the quality of a fishery in terms of species composition and species richness declines. Eutrophication can negatively affect the types of fish. In the case of drinking water supplies, eutrophication can increase water treatment costs, contribute to taste and odor problems, and increase production of trihalomethanes during the treatment process. Accelerated eutrophication also results in blue-green algal blooms that make lake water unpalatable and possibly unsafe for drinking water and may limit aquatic recreational activities.

2.3.2 Significant Issues

2.3.2.1 Watershed Context

Lake nutrient concentrations are determined by the geological characteristics of the lake's watershed and lake morphometry (Rawson 1939). Lake data show that catchment geology, soil fertility, and important landform features such as catchment area/lake area ratios and lake depth largely determine lake trophic state (Duarte and Kalff 1989). In a given geologic setting, deep lakes are less productive than shallow ones. Simply, the edaphic condition establishes the "potential" trophic state of a particular lake. Agriculture and urbanization contribute to eutrophication and cause waterbodies to become more eutrophic through nutrient loading in excess of inherent background levels set by the edaphic situation. Lakes in fertile lands, such as the midwestern United States, are characteristically eutrophic, in contrast to the naturally oligotrophic lakes located within comparatively infertile landscapes, such the granitic Canadian Shield or Rocky Mountains. This spatial pattern (Omernik et al., 1991) follows the general soil fertility of the terrestrial landscape and is the basis for the ecoregion approach for setting nutrient criteria (U.S. EPA 1998).

2.3.2.2 Internal Nutrient Cycling

Complex physical, chemical, and biological processes determine the internal distribution and fate of nutrients introduced into aquatic systems. The net efficiency of internal nutrient cycling is governed by physical mixing and water turnover time, aquatic chemical reactions that influence the dissolution, speciation, and bioavailability of nutrients, and finally, aquatic food web dynamics.

A detailed discussion of the extensive technical literature describing internal nutrient cycling relevant to lakes and reservoirs is beyond the scope of this report. Discussions of the key aspects regarding nutrient cycling in lakes can be found in limnological (Wetzel 1983) and ecological (Odum 1971) texts. More recent examinations concerning the importance of sediments as internal nutrient sources and the role of trophic interactions in regulating nutrient cycling are briefly described in the following sections.

2.3.2.3 Sediment Nutrient Inputs

The sediments of eutrophic lakes are often an important source of P (and other potentially limiting nutrients) to the overlying water. The redox chemistry of iron in relation to DO depletion has been demonstrated as a chemical mechanism that remobilizes P from sediments into the water column. Additionally, the role of benthic consumer organisms in the release of inorganic P into the water column has been known for some time (Kitchell et al. 1979). Physical disturbance of sediments by benthic fishes has also been suggested as a mechanism for adding nutrients back into the water column. Many shallow lakes have no history of high external loading but nevertheless experience high maximum summer phosphorus values and corresponding large algal blooms. Releases of nutrients from sediments can explain these algal blooms.

The migration of P-laden cells from the lake bottom to the water column as an internal loading mechanism has been supported by field evidence (Barbiero and Welch 1992, Barbiero and Kann 1994, Welch and Cooke 1995, Head et al. 1999). Several shallow lakes have been shown to experience episodes of significant internal loading coincident with the development of large planktonic blue-green populations, suggesting that the increase in P is the result of translocation from the sediments via algal migration.

The potential for nutrient-laden sediments to serve as the primary source of nutrient inputs to lakes carries important implications for the development of SSNC. Implementation of nutrient criteria for lakes dominated by inputs of nutrients from sediments may prove impossible or require prohibitively expensive dredging and disposal programs.

2.3.2.4 Trophic Cascade

The impacts of nutrient enrichment on biological response variables (phytoplankton biomass, for example) are influenced by the food web interactions characteristic of a particular lake or reservoir. Research continues to focus on the relative importance of primary producers and consumer organisms in controlling the distribution of biomass across different trophic levels in aquatic systems. Under certain circumstances such as the late spring and early summer pulse of primary production, the distribution of biomass is determined by the ability of phytoplankton to assimilate available nutrients and grow rapidly. This new production, assimilated through successive trophic levels via grazing and predation, determines the population sizes of zooplankton, other invertebrates, and fish. The system is structured from "the bottom up" from a trophic perspective (Bartell et al. 1989). Alternatively, predator-prey relationships can structure the food web from "the top down," and the abundance of phytoplankton is determined by the alterations in zooplankton grazing that result indirectly from changes in food web interactions among fish (piscivores and planktivores). Importantly, the relative dominance of top-down versus bottom-up structuring of aquatic food webs can alternate through time.

Understanding this trophic cascade (Carpenter and Kitchell 1993) in lakes and reservoirs can be crucial when establishing effective SSNC. If, for example, phytoplankton biomass is determined primarily by top-down trophic interactions, reducing (or increasing) the availability of nutrients in that system will not likely result in measurable changes in phytoplankton.

2.4 Estuaries

2.4.1 Overview

Estuaries are highly dynamic and exceedingly productive. The ecological characteristics of estuaries are strongly influenced by the physical mixing of freshwater riverine inputs, stratification, and tidally driven saltwater movements. The dynamic physical and chemical nature of estuarine systems can produce highly variable and scale-dependent biological and ecological conditions. As a result of this characteristic variability, it may prove extremely difficult to derive or implement SSNC for these heterogeneous systems. Given the location of estuaries at the downstream end of complex drainage networks, the development and application of SSNC for upstream lakes, rivers, and reservoirs can influence SSNC for estuaries. Alternatively, the deviation and implementation of SSNC in estuaries may influence or constrain the development of SSNC in upstream lakes, reservoirs, and rivers.

Estuaries are typically highly productive and critical habitats for the life stages of important marine species. Additionally, these systems are important foraging habitats for important coastal and marine resources (crabs, oysters, and striped bass and other fish). Therefore, while perhaps prohibitively difficult in the end, serious efforts should be devoted in concept and method to the development of SSNC consistent with desired use and use attainability for these critically important ecosystems.

2.4.2 Significant Issues

The derivation of SSNC for estuaries may reasonably involve issues previously identified in the development of regional criteria. For example, the U.S. EPA identified the following research topics that would facilitate derivation of regional nutrient criteria for estuaries and coastal marine systems (U.S. EPA 1998):

- nitrogen and phosphorus limitation in relation to salinity gradients;
- role of DO in excessive nutrient enrichment;
- excessive enrichment, sedimentation, and turbidity;
- biological community indicators;
- relative importance of tidal and freshwater (watershed) influxes of nutrients;
- flushing rates;
- impact of shore discharges on estuaries; and
- harmful algal blooms.

These topic areas highlight potentially important information needs that may have to be addressed on a site-specific basis in the derivation of meaningful nutrient criteria for estuaries. Three topic areas are discussed below.

2.4.2.1 Tidal Wedge or Vertical Density Stratification

Seasonal variations in precipitation combine with varying tidal strengths to produce complex and dynamic patterns of water circulation and mixing of freshwater and saltwater within estuaries. The patterns of water circulation in nearly all the estuaries in the United States are influenced by tides. The combination of downstream flows of freshwater and upstream movement of tidally driven, denser saltwater results in the formation of a saltwater wedge. Most estuaries have relatively diffuse salinity boundaries at their landward end. Tides, freshwater flow, vertical heat gradients, and wind-driven water levels associated with bottom topography all determine the nature of the longitudinal salinity gradient and other physical water mixing and circulation features. The location and stability of this ecotone is determined by the relative magnitudes of freshwater flow and the strength of the tides, as well as the bathymetry of the estuary. Physical mixing across the tidal wedge can influence nutrient concentrations in both the freshwater and underlying layer of saltwater.

The establishment of effective SSNC for estuaries will require a quantitative characterization of the seasonal and spatial patterns of physical circulation and mixing of freshwater and saltwater for the estuary of interest. The characterization of these patterns will be crucial in the definition of what is meant by "site" in the derivation of site-specific nutrient criteria for individual estuaries.

2.4.2.2 Nutrient Import/Export

Tidal mixing raises the possibility that estuaries may receive nutrient inputs from the coastal marine environment. Alternatively, periods of dominant downstream flows of freshwater may result in the export of nutrients and organic matter to coastal ecosystems. Identifying those conditions of physical mixing and biological productivity conducive to an annual net import or net export of nutrients remains a focal area of research in estuarine ecology. The derivation of SSNC and the effectiveness of their implementation could be strongly influenced by the tendency for an estuary to be a net importer of nutrients from surrounding coastal waters.

The combination of spatial-temporal variability in the physical circulation patterns within the estuary (flushing) and nutrient inputs (point source and nonpoint source) from the surrounding watershed and adjacent coastal ocean (external loading) may require the derivation of several SSNC that apply at different times and locations. The critiera may be loading based rather than concentration based. The difficulty in deriving and putting into practice varying SSNC will challenge current site-specific understanding of most estuarine systems and may preclude these systems from meaningful (and enforceable) SSNC.

2.4.2.3 Microbial Loop

The description and discussion of the microbial loop presented previously for lakes and reservoirs is similarly relevant to the derivation of SSNC for estuaries. The relationship between nutrients and the production dynamics of the loop takes on additional importance in estuaries because of the potential for the development of harmful algal blooms (toxic dinoflagellates, for example) and reduced energy transfer to desirable higher trophic organisms.

2.5 Tools for Developing Data for Site-specific Nutrient Criteria

The derivation of SSNC is fundamentally a scientific enterprise, and the available tools include many of those used in basic and applied aquatic ecology. The following sections outline several methods that contribute to increasing the quantitative understanding of the structure and function of aquatic ecosystems. Logically, SSNC derive from this understanding and involve the use of many of these tools.

2.5.1 *In situ* Experiments

The quantitative understanding of responses to nutrient enrichment is based on a history of experiments performed directly within aquatic ecosystems. These experiments range from the artificial enrichment and *in situ* incubation of 300 mL glass bottles containing resident plankton, to enclosed portions of lakes (limnocorrals), to nutrient amendments applied to extremely large marine enclosures (CEPEX), and finally whole-lake manipulations (Schindler 1990).

Similar experimentation may be used to develop the quantitative basis underlying the derivation of SSNC. The clear advantage of this approach is that definite site-specific relationships between nutrient enrichment and system response are obtained. Experiments can be designed and performed to focus directly on the ecological resources or water quality objectives of interest. The disadvantages include the possible need to perform many experiments to examine system responses over the requisite spatial-temporal scales to derive nutrient criteria.

2.5.2 Field Monitoring

If controlled experiments prove an infeasible approach in developing site-specific criteria, data produced by carefully designed field monitoring studies may provide the inferential relationships between nutrient loading and site-specific environmental responses that can serve as the basis for deriving the desired criteria. Data obtained from previous or ongoing monitoring studies may have to be used if time or resources do not permit a planned field monitoring program.

2.5.3 Empirical Nutrient-loading Models

During the past three decades, the concepts of trophic state and the eutrophication process have been bolstered by quantitative relationships that link lake nutrient content to its external load through mass-balance calculations (Vollenweider 1968). Substantial evidence demonstrates that lake trophic state is determined by in-lake P and N values (Schindler 1977, Downing and McCauley 1992). Across a continuum from low to high nutrient levels, potential response variables (algal biomass, bacterial numbers, and turbidity) increase with increasing nutrient concentrations. The technical literature includes several empirical models that quantify these biological responses to lake nutrient levels (Smith 1998). These models may provide a quantitative basis for deriving site-specific criteria and subsequent management of aquatic ecosystems.

2.5.4 Ecological Process Models

In contrast to the kinds of statistical nutrient loading models described above, ecological process models attempt to describe, in mathematical formulations, the complex physical, chemical, biological, and ecological phenomena that determine the transport, fate, and effects of nutrients added to aquatic ecosystems. A comprehensive and detailed history of the development of ecological process models lies beyond the scope of this report. However, several of the early aquatic ecosystem process models (Patten et al. 1975; Park et al. 1974; MacCormick et al. 1975) were developed during the 1960s and 1970s as part of the International Biological Program (IBP) (McIntosh 1985). These models were characterized by dozens of chemical and biological state variables, hundreds of parameters, and nontrivial requirements of environmental data (light intensity, water temperature, and nutrient loading rates) to perform the model computations. The aquatic process models developed in support of the IBP emphasized the quantitative characterization of ecosystem productivity. Other aquatic ecosystem models developed independently of, or subsequent to, the IBP were similar in form and function. Other models, such as the Hydrologic Simulation Program-Fortran (http://www.epa.gov/ceampubl/swater/index.htm) and the Spatially Referenced Regressions on Watershed Attributes model (SPARROW, http://water.usgs.gov/nawqa/sparrow) are watershed models driven by loading parameters. These models explicitly simulated the impacts of nutrient enrichment (or limitation) on changes in algal biomass; water quality parameters (suspended particulate matter, DO); benthic productivity; and fish population sizes. Thus, many if not all of the impacts relevant to establishing SSNC were addressed by these complex models. Therefore, ecological process models may prove to be effective tools in the derivation of SSNC.

The utility of complex ecological models in establishing SSNC will be determined by factors including the following:

- availability of models appropriate to the ecosystem of interest,
- quality and quantity of necessary site-specific data,
- demonstrated accuracy and reliability of previous model applications,
- time and economic resources for site-specific model implementation,
- ability to translate model results to permissible nutrient loading, and
- ability to modify and refine the model in relation to site-specific monitoring.

While perhaps initially daunting, each of these factors can be effectively addressed. A diverse set of aquatic ecosystem models has accumulated over nearly 30 years of modeling. Site-specific data can be developed or substituted with data from similar or nearby systems, at least for preliminary simulations. The resources needed for model implementation and execution can be comparatively economical, especially when weighed against the high costs of field experiments and monitoring. Issues of model accuracy and reliability must include consideration of requirements for sound decision making.

The efficacy of aquatic ecosystem models has been demonstrated in the characterization of the impacts of nutrient enrichment (Park et al. 1974; MacCormick et al. 1975; DiToro et al. 1975), as well as in the estimation of ecological risks posed by toxic chemicals (Bartell 1994; Bartell et al. 1992, 1999, 2000; Park 1998). This modeling approach may prove effective in developing SSNC, particularly for Tier 3 evaluations.

CHAPTER 3.0

TIER 1 METHODS FOR DEVELOPING SITE-SPECIFIC NUTRIENT CRITERIA

3.1 Introduction

This chapter describes each step in the Tier 1 SSNC process. In Tier 1, SSNC are developed using available and/or new data and relatively simple data analyses. Therefore, the cost of developing Tier 1 criteria should be relatively low. Finally, management plans are outlined to implement the criteria and evaluate their success. The Neuse River in eastern North Carolina is used as a case study to illustrate Tier 1 methods.

3.2 Problem Description

The Neuse River is a eutrophic, fourth- and fifth-order coastal plain river located in the Southeastern Temperate Forested Plains and Hills Ecoregion (U.S. EPA 1998). The source of the Neuse River is north of Durham in the Piedmont region of North Carolina. The river flows approximately 320 km through the Piedmont to the Coastal Plain and discharges into the southern end of Pamlico Sound. The case study focuses on a segment of the river near Kinston.

The Neuse River near Kinston is classified as Class C (freshwater aquatic life, secondary recreation) and nutrient sensitive water (NSW). All designated uses of this segment of the Neuse River, including its aquatic life, secondary recreation, and status as a nutrient sensitive water, are potentially affected by nutrients. The nutrient water quality objective for this segment of the Neuse River is assumed to be attainment of these designated uses. Currently, the only numerical nutrient criterion for the Neuse River near Kinston is the statewide Chl *a* criterion, 30 µg/L; an additional requirement is that nutrients cannot increase above background levels.

The Neuse River near Kinston is located in U.S. EPA Ecoregion 14, Level III Ecoregion 63 (or subregion). The U.S. EPA (2000c) estimated Ecoregion XIV 25th percentile TN, TP, and seston Chl *a* to be 710 µg/L, 31 µg/L, and 3.8 µg/L, respectively (Table 3-1). For Level III Ecoregion 63, the values were 870 µg/L for TN, 53 µg/L for TP, and 3.8 µg/L for seston CHL *a*. For Ecoregion XIV no Chl *a* values for periphyton were published (U.S. EPA 2000c). U.S. EPA's (2000b) guidance states that these 25th percentile nutrient and biomass concentrations can be used as default ecoregional nutrient criteria for the Neuse River.

Table 3-1. Ecoregional 25th Percentile Values for TN, TP, Chl *a*, and Total Suspended Solids (TSS)

Source	TN (µg/L)	TP (µg/L)	Seston Chl *a* (µg/L)	TSS (mg/L)
U.S. EPA Ecoregion XIV Streams	710	31	3.8	Not available
U.S. EPA Level III Ecoregion 40 Streams	870	53	3.8	Not available

Designated uses appear to be attained in this segment of the Neuse River. However, downstream in the Neuse River Estuary, which is part of Pamlico Sound, the river exhibits many symptoms of eutrophication, including excessive algal growth, bottom-water hypoxia, and fish kills. These symptoms suggest that the designated use for aquatic life related to nutrients is not being attained downstream of Kinston in the Neuse Estuary. In the case study, the default ecoregional criteria are not met; however, stakeholders and the interested public in and around Kinston know that the designated uses of the river are. Thus, the decision is whether less stringent site-specific nutrient standards should be set for this segment of the Neuse River. According to the decision matrix presented in Chapter 1.0, Table 1-3, when designated uses are met, but criteria are exceeded, the regional nutrient criteria may be over protective of the desired uses or objectives, and less stringent site-specific nutrient criteria may be justified.

In the beginning of the SSNC process, the nutrient issues to be addressed by the SSNC should be thoroughly described, especially as they relate to uses of the waterbody and to upstream and downstream waterbodies and their uses. Initially, this can be done by formulating a series of questions and preliminary answers. For the Neuse River at Kinston these included the following:

- Is there a nutrient-related water quality problem in the waterbody? —No.
- Do U.S. EPA ecoregional nutrient criteria appear overly stringent or insufficiently stringent to protect designated or other uses? —The ecoregional criteria may be more stringent than necessary.
- Are uses of the waterbody adversely affected by nutrients? —No.
- What are the uses apparently affected by nutrients? —None.
- How are nutrient problems manifested? —High levels of nitrate are present in the river and may be affecting downstream users.
- What nutrients are causing or associated with water quality problems? —Unknown.
- What are the apparent sources of the nutrients? —External, point and nonpoint sources, including concentrated animal feeding operations, municipal wastewater treatment plant effluents, and runoff.
- How is the suitability of nutrient criteria determined? —Concentration of nutrients and designated uses.
- Do upstream and/or downstream stream segments or waterbodies have nutrient problems? —The primary problem areas are in the downstream estuary.

- Will upstream or downstream waterbodies or segments be adversely affected by SSNC? —The Neuse Estuary could be adversely affected by less stringent SSNC in the Kinston segment of the river.
- Are there other non-nutrient issues that could be contributing to these problems? —Overfishing maybe an issue in the Neuse River estuary.
- Will stakeholders support SSNC? —Less stringent SSNC may be controversial because of concerns for the Neuse River.
- Are resources available to support the studies required to develop defensible SSNC? —Yes.

3.3 Selecting Important Variables

The primary nutrient variables that are candidates for use in developing Tier 1 SSNC are TN and TP. The primary biological response variables are either periphyton or phytoplankton biomass measured as Chl *a*, and a measure of water clarity, such as TSS or turbidity. DO and SAV biomass may be important response variables in estuaries, deep lakes that stratify, and deep, slow rivers. The primary nutrients (TN and TP) and biological response variables (Chl *a* and water clarity) should be components of all SSNC. Additional variables may be of interest, especially in proceeding to Tiers 2 and 3 (Table 3-2). For the Neuse River, TN and TP were the nutrient variables and Chl *a* and TSS the response variables selected for Tier 1 analysis. In addition, downstream estuarine effects should be considered using DO and the presence of nuisance algal blooms as metrics.

Table 3-2. Variables that May be Important in Developing SSNC

Nutrient Variables
♦ total phosphorus
♦ total nitrogen
♦ dissolved inorganic nitrogen
♦ soluble reactive phosphorous
Primary Response Variables
♦ chlorophyll *a*
♦ water clarity
♦ dissolved oxygen
♦ volatile suspended solids
Secondary Response Variables
♦ anoxic factor
♦ areal hypolimnetic oxygen depletion rates
♦ benthic macroinvertebrate community metrics
♦ zooplankton community metrics
♦ grazing
♦ fish population and community metrics
♦ SAV
♦ algal community metrics
Other Potentially Important Variables
♦ discharge/velocity
♦ substrate
♦ point and nonpoint inputs
♦ basin morphology and hydrology
♦ longitudinal gradients
♦ flushing rate
♦ catchment area/lake area ratios

3.4 Establishing an Appropriate Data Base

The derivation of Tier 1 SSNC emphasizes the use of existing data, although additional data collection may be required. Data describing both nutrient concentrations and loading rates are critical. Loading data are especially useful for lakes, reservoirs, and estuaries because sampling cannot be conducted on time/space scales for every waterbody to account for variability in nutrient concentrations. Relying solely on concentration data may not lead to successful SSNC. Data quantifying TN, TP, Chl *a*, and water clarity are required for periods of peak algal biomass or during other periods when algal problems can occur. Data for DIN and SRP, secondary response variables, and other variables that influence primary production and algal biomass can be valuable in some Tier 1 analyses. However, these data will most likely be more useful for the more intensive Tier 2 or 3 analysis. For Tier 1, sufficient data are required to characterize the spatial and temporal variability of each variable being evaluated. Data should include mean values, minimum and maximum values, and variances. Data for at least one entire year based on bimonthly or more frequent sampling are recommended. Data for a single year may be adequate, if the year had relatively typical or average climatic and hydrological conditions. If the year is atypical, with abnormally wet, dry, warm, or cold conditions, several years of data may be required.

Table 3-3 lists sources of nutrient and other water quality data that can be used to develop SSNC. Additional data and information may be available from state environmental agencies, universities, and county or local government organizations.

Table 3-3. Examples of Data Sources for Use in Developing Site-specific Nutrient and Water Quality Information

Organization	Description	Website
U.S. EPA	Nutrient Criteria Database	http://www.epa.gov/waterscience/criteria/nutrient/database/index.html
U.S. EPA	STORET - data storage and retrieval system	http://epa.gov/storet/
USGS[1]	National Water Information System – NWIS	http://water.usgs.gov/nwis
USGS/U.S. EPA	National Hydrography Dataset	http://nhd.usgs.gov/
U.S. EPA	Watershed Information Network – WIN	http://epa.gov/win/
U.S. EPA	Surf Your Watershed	http://epa.gov/surf/

[1]USGS: U.S. Geological Survey

In some parts of the United States, nutrient data are available in a readily useable format. Examples of good lake survey information databases are Iowa (Jones and Bachmann 1978), Texas (Ground and Groeger 1994), Minnesota (Heiskary and Walker 1988), Missouri (Jones and Knowlton 1993), Colorado (Morris and Lewis 1988), and Florida (Canfield and Hoyer 1988). Compilations of lake and reservoir data are available for some regions (Reckhow 1988). In many regions, however, considerable water quality information is available only for a small number of economically important waterbodies, and data for the remaining systems are scant. Thus, characterizing site-specific water quality from existing data could prove difficult.

Nutrient data are temporally variable (Knowlton et al. 1984) within and between years. One potential problem with existing data is the temporal variation can confound the derivation of cases to response relationships. Evaluations using such data can be improved by using long-term averages (Jones et al. 1998). Analysis of water quality from Missouri reservoirs, for example, shows that at least three years of data are required to account for such variance and to adequately characterize trophic state characteristics of individual reservoirs.

For many waterbodies, available data are likely to be inadequate to develop defensible SSNC. Of the two primary nutrient variables (TN and TP) and the response variables (Chl *a* and water clarity) data for TN, TP, and water clarity are available for many waterbodies, but data for Chl *a* may only be available for a relatively few. New data may have to be collected before SSNC can be developed. The U.S. EPA (2000a, 2000b, 2001) provides guidance on sampling plan designs for nutrient studies in lakes and reservoirs, rivers and streams, and estuaries.

3.5 Data Analysis

For the Neuse River, data from 1975-1995 were compiled for nutrients and primary and secondary response variables from stations located throughout the watershed. These data were provided by Craig Stowe, Duke University, Durham, N.C. The first step was to plot the data and look for general trends and relationships. Figure 3-1 summarizes trends in TN, TP, Chl *a,* and TSS in the Neuse River near Kinston from 1982-1995. From 1982-1995, TN concentrations remained fairly consistent. Beginning in 1988, however, TP concentrations declined when detergents containing phosphorous were banned from use in the watershed. Chl *a* levels also decreased during this period, suggesting that it may have responded to the decline in TP. In the Neuse River at Kinston, analysis of the data compiled from 1982–1995 indicates that TP averaged 136 μg/L, TN averaged 1,481 μg/L, and the N:P ratio of the averages is 10.9. These observations suggest that this segment of the river is phosphorus limited.

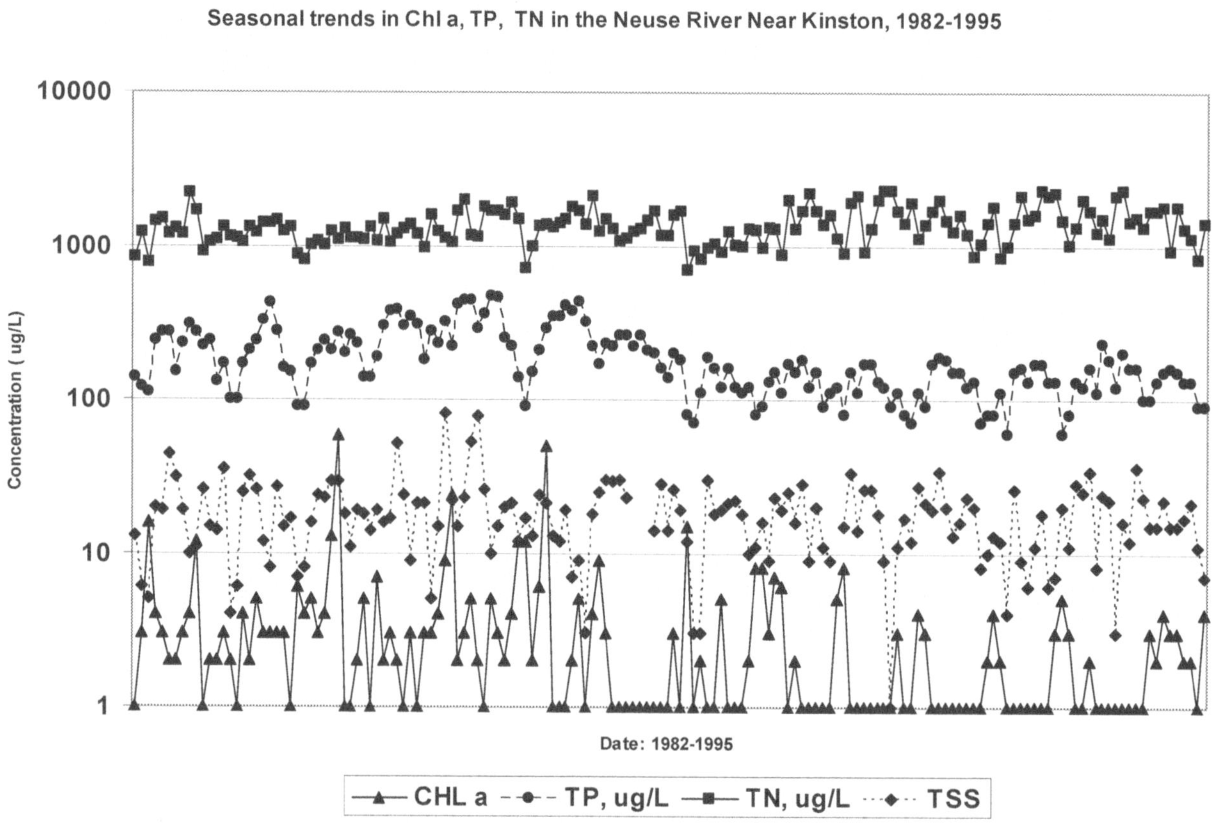

Figure 3-1. Seasonal Trends in Chl *a*, TP, TN, and TSS in the Neuse River near Kinston, NC, 1982-1995

The second step in Tier 1 data analysis is to create scatter plots for Chl *a* versus TN and TP to determine if any apparent relationships are present between the nutrients and response variables. Generally, these plots are done on a Base 10 logarithmic scale or on $\log_{10}$-transformed data. The following comparisons can be done to define relationships between nutrients and algal biomass:

- measurements of Chl *a*, TN, and TP among seasons;
- growing season comparisons of Chl *a*, TN, and TP;
- summer mean total limiting nutrient concentration with summer mean and maximum Chl *a*;

- pre-maximum growth period (spring, pre-runoff) mean dissolved limiting nutrient concentration with maximum algal biomass;
- mean annual dissolved nutrient concentration with the 75th percentile mean algal biomass;
- cellular concentrations of the limiting nutrient with maximum algal biomass.

Figure 3-2 compares mean monthly values of Chl *a* and TN from 1982-1995. Significant relationships between Chl *a* and TN apparently do not exist. The same appears true for Chl *a* and TP (Figure 3-3).

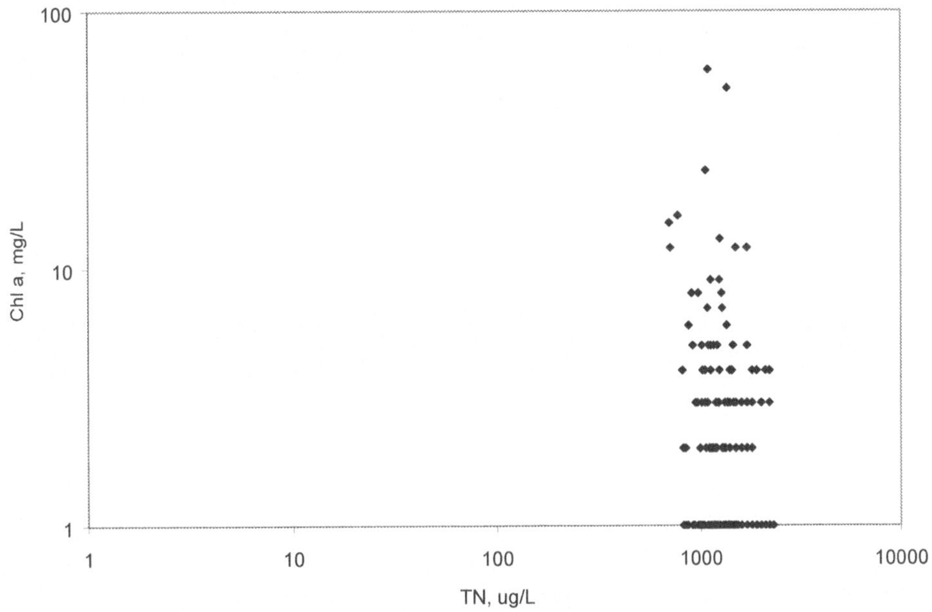

Figure 3-2. Comparison of Chl *a* Versus TN in the Neuse River near Kinston, NC, 1982-1995

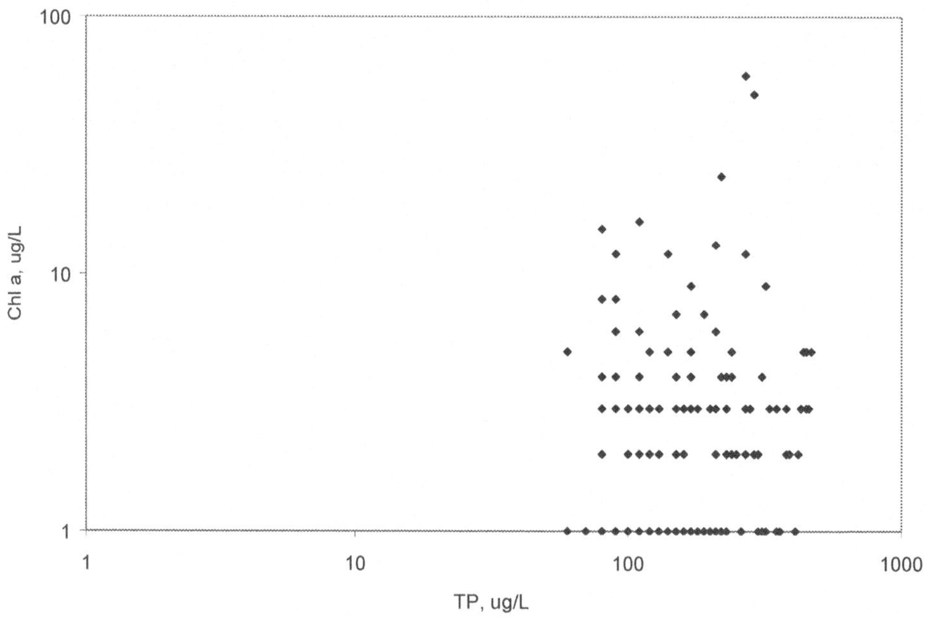

Figure 3-3. Comparison of Chl *a* Versus TP the Neuse River near Kinston, NC, 1982-1995

Other factors such as color, turbidity, or TSS appear to confound the relationships between TN and Chl *a* and TP and Chl *a*. Much of the Neuse River has low transparency because of naturally high concentrations of dissolved organic carbon, and the river is extensively shaded due to riparian tree cover. Therefore, photosynthesis may be inhibited by lack of light, with algal biomass correspondingly reduced. The only measure of water transparency available, however, was TSS, and no relationships between TSS and Chl *a* were apparent (Figure 3-4). The lack of significant relationships between TN, TP, and Chl *a* in this segment of the Neuse River suggests that efforts to reduce nutrient concentrations to meet U.S. EPA ecoregional nutrient criteria most likely would not reduce Chl *a*. The relatively low Chl *a* concentrations and relatively high TN and TP concentrations support the development of less stringent Tier 1 SSNC for the Neuse River at Kinston.

The third step in Tier 1 data analysis generally involves developing cumulative frequency distributions (CFDs) of the data for each nutrient and biological response variable. These distributions are developed by ranking variable values from lowest to highest, calculating the cumulative percentiles for each value, and then plotting them. CFDs provide information on the proportion of samples that match (or exceed) a selected criterion.

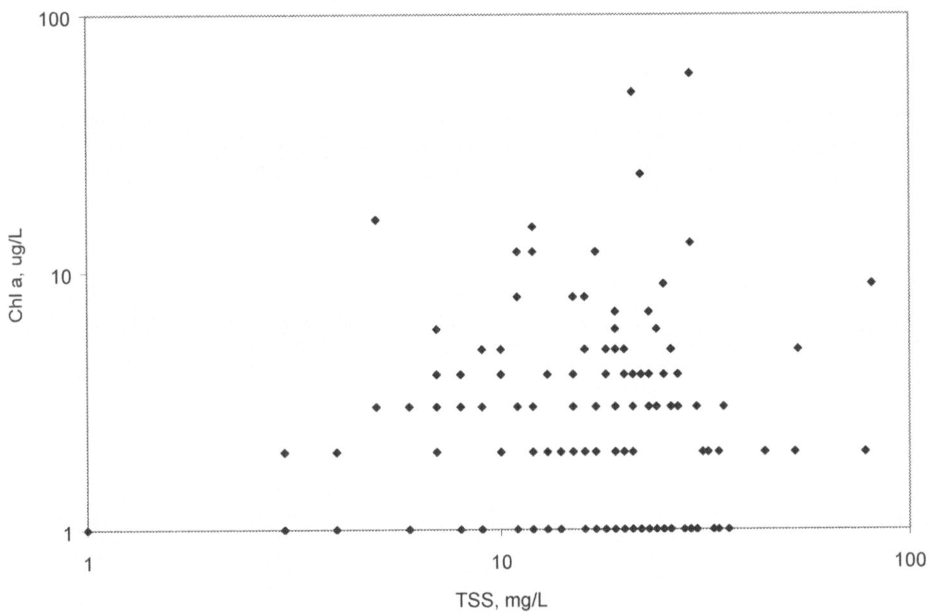

Figure 3-4. Comparison of Chl *a* Versus TSS in the Neuse River near Kinston, NC, 1982-1995

Figures 3-5 through 3-8 are cumulative frequency distributions for measurements of TN, TP, seston Chl *a*, and TSS in the Neuse River at Kinston from 1982 to 1995. The lower 25th percentile values are 1,110 µg/L, 125 µg/L, 1.6 µg/L, and 12 mg/L, respectively. These or other values could be selected as Tier 1 SSNC if they were found to be associated with acceptable water quality conditions such as decreased algal blooms and fewer fish kills and did not adversely affect downstream uses. The site-specific 25th percentile values for TN and TP for the Neuse River at Kinston were greater than both the Ecoregion IX and Level III Ecoregion 40 values (Table 3-1), while the site-specific 25th percentile value for Chl *a* was much less than the ecoregional value. This inconsistency can be expected when comparing regional values to site-specific measurements.

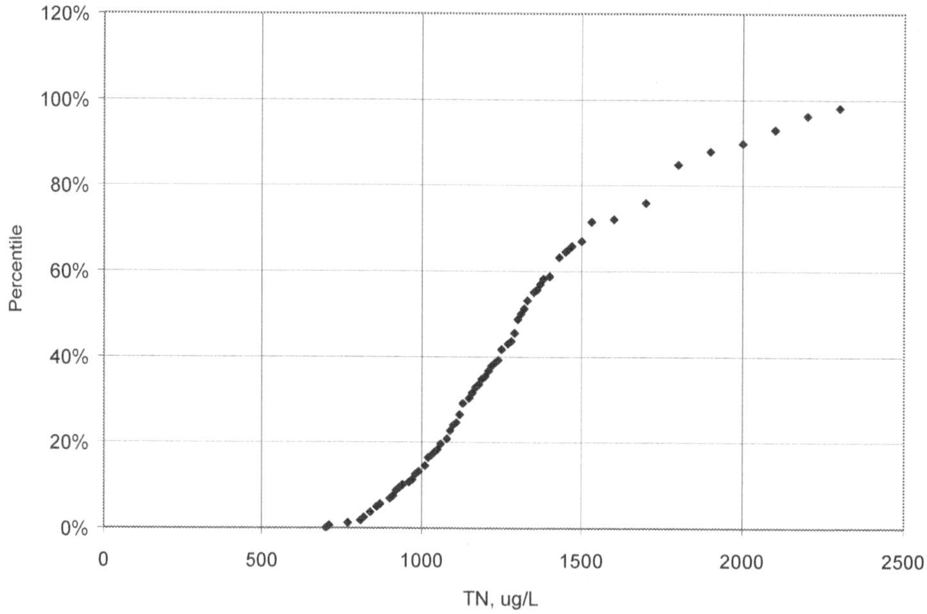

Figure 3.5. Cumulative Distribution of TN in the Neuse River near Kinston, NC, 1982-1995

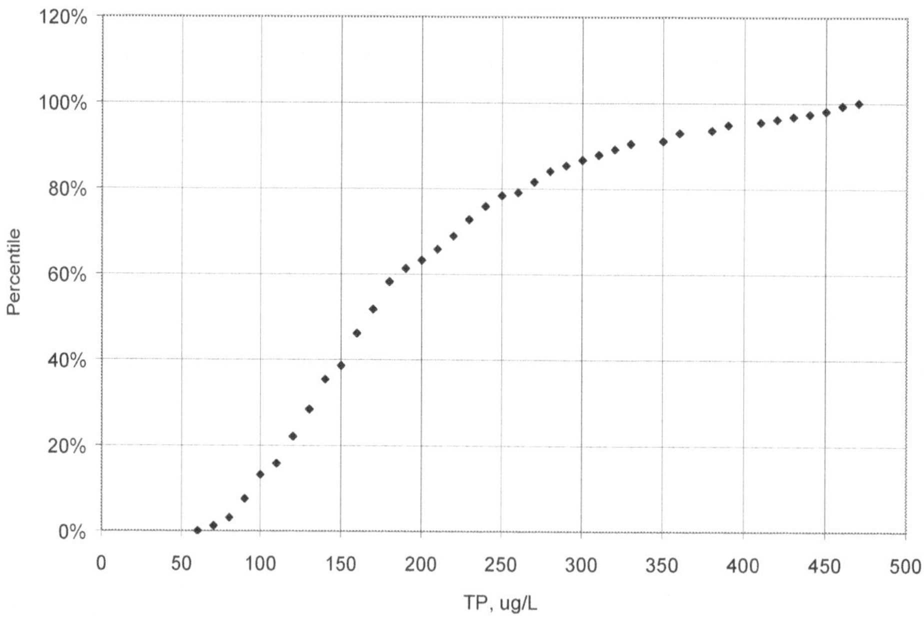

Figure 3-6. Cumulative Distribution of TP in the Neuse River Near Kinston, NC, 1982-1995.

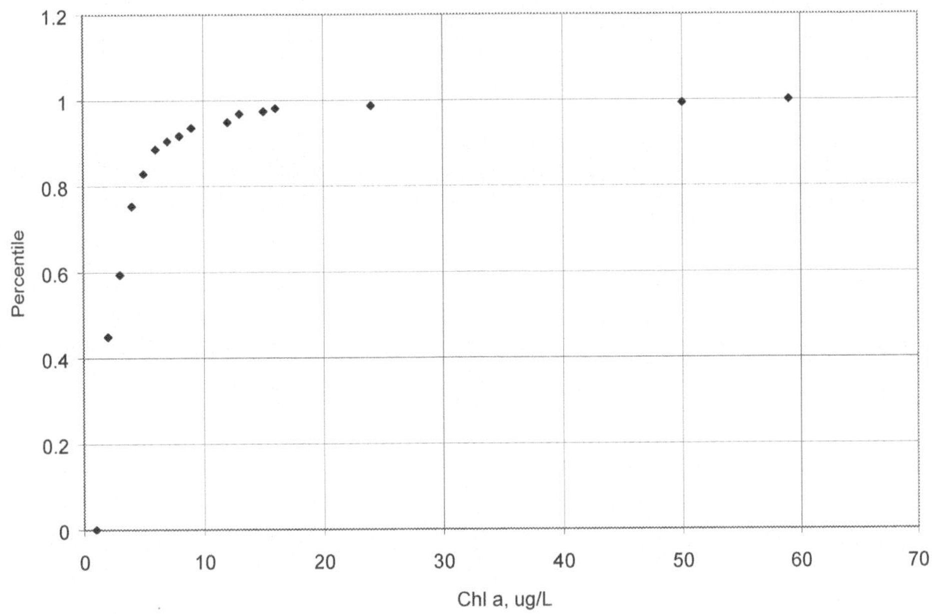

Figure 3-7. Cumulative Distribution of Chl *a* in the Neuse River Near Kinston, NC, 1982–1995.

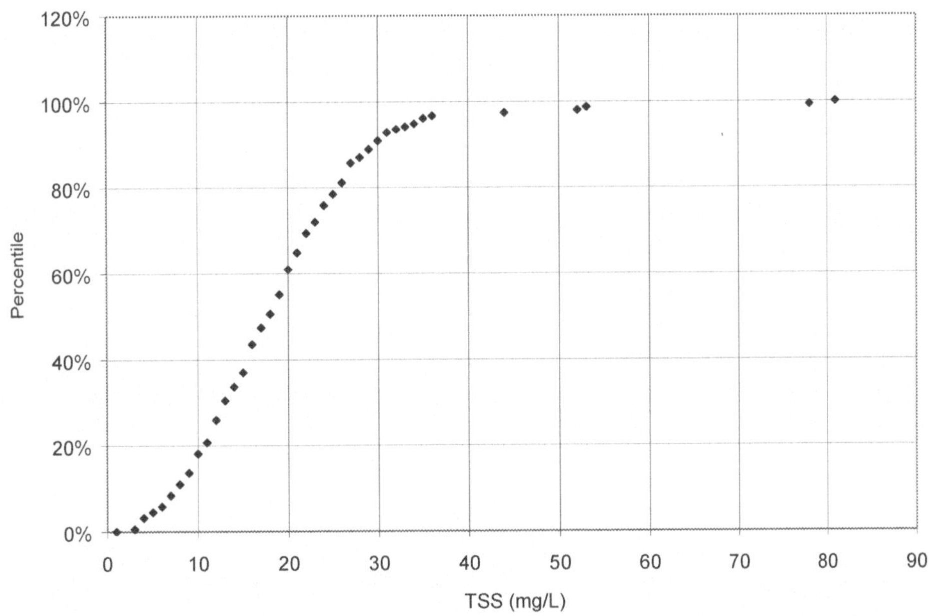

Figure 3-8. Cumulative Distribution of TSS in the Neuse River near Kinston, NC, 1982-1995

3.6 Selection of Tier 1 Site-specific Criteria

Tier 1 SSNC can be developed in the same manner that U.S. EPA (2000a, 2000b, 2001) develops its ecoregional criteria. The agency recommends that the ecoregional nutrient criteria be estimated from the lower 25th percentile of the CFD of all data for each nutrient or the upper 75th percentile of the CFD of data from reference waterbodies. Similarly, a potential SSNC can be estimated from the lower 25th percentile, or some other proposed percentile, of all data for each nutrient from the waterbody of interest, if the waterbody is not a reference waterbody. If the particular system is a reference waterbody, the upper 75th percentile can be used as the proposed SSNC.

Tier 1 SSNC can also be selected using other percentiles of the CFDs for TN and TP, for example concentrations associated with acceptable Chl *a* values. Acceptable Chl *a* values are those associated with acceptable water quality conditions and designated uses. An analysis can be done to determine whether designated uses potentially affected by nutrients are being attained and if the cause of the nonattainment is directly or indirectly related to excess nutrients. Information on methods for conducting these analyses for rivers and streams is found in U.S. EPA (1983) and Water Environment Research Foundation (WERF) (1995) documents. The designated uses of interest for this example analysis are those potentially affected by nutrients, such as aquatic life, fisheries, recreation, swimming, or drinking water supply. For example, high levels of algae, low DO concentrations, and/or high pH levels caused by high concentrations of algae may adversely affect the types of aquatic species present and may cause or contribute to fish kills. These effects may prevent attainment of the designated use for aquatic life or fisheries. Floating or attached algae may interfere with recreational fishing and swimming. Algae may cause unacceptable taste and odor problems in a river, lake, or reservoir used as a source of drinking water. Excess nutrients may also contribute to exceedances of DO and pH standards.

Another approach that can be used to quantify perception of river or lake use for recreation and aesthetics is a stakeholder survey. (Appendix A describes a proposed framework for stakeholder involvement in the SSNC process.) Such a survey would compare stakeholder perceptions of water quality and stakeholder uses of the waterbody with measurements of TP, TN, Chl *a*, water clarity, and other nutrient and response variables. Then, relationships between use perception and nutrient-related water quality would be determined. The results of such a study could be used to derive site-specific nutrient-related water quality conditions acceptable to stakeholders. Tables 3-4 and 3-5 present results of an example survey for a hypothetical lake SSNC derivation.

Table 3-4. Hypothetical Lake Survey Part A

A. Describe condition of lake today		TN (ppb)	TP (ppb)	Secchi Depth (ft)	Chl *a* (ppb)
1	Crystal clear	30	15	10	3
2	Some algae visible	75	25	7	7
3	Green color	150	45	4	12
4	Limited clarity, mild odor	250	80	2	22
5	Floating scums, foul odor, dead fish	300	120	0.5	35

Table 3-5. Hypothetical Lake Survey Part B

B. Describe suitability of lake for recreation		TN (ppb)	TP (ppb)	Secchi Depth (ft)	Chl *a* (ppb)
1	Beautiful, couldn't be nicer	30	15	10	3
2	Minor aesthetic problems, excellent for enjoyment	75	25	7	7
3	Swimming and aesthetics slightly impaired by algae	150	45	4	12
4	Swimming & aesthetics severely impaired by algae	250	80	2	22
5	Swimming & other recreation nearly impossible	300	120	0.5	35

The literature shows that user perception differs regionally, and it probably also will differ on a site-specific basis. One of the best examples is from Minnesota, where large regional differences occur in the concentration of P (an indicator of lake trophic state) considered acceptable for drinking water supply and primary contact recreation (Table 3-6). Perception regarding the level of algal biomass that constitutes a nuisance condition (Table 3-7) also varies across the region. In Oregon, Chl *a* levels > 15 µg/L are considered to impair the beneficial uses of natural lakes, reservoirs, rivers, and estuaries, whereas in North Carolina, the Chl *a* criterion is 30 µg/L. Chl *a* criteria for Ontario lakes are quite restrictive relative to other regions in North America (Table 3-8). These differences reflect regional water quality objectives and related uses that are locally attainable.

Table 3-6. Most Sensitive Lake Uses by Ecoregion and Corresponding Phosphorus Criteria for Minnesota (Heiskary and Walker 1988)

Ecoregion	Most Sensitive Uses	P Criteria (µg/L)
Northern Lakes and Forests	Drinking water supply	<15
	Cold water fishery	<15
	Primary contact recreation and aesthetics	<30
North Central Hardwood Forests	Drinking water supply	<30
	Primary contact recreation and aesthetics	<40
Western Corn Belt Plains	Drinking water supply	<40
	Primary contact recreation and aesthetics (full support) (partial support)	 <40 <90
Northern Glaciated Plains	Recreation and aesthetics (partial support)	<90

Table 3-7. Aesthetic or Use Impairment Classification Systems Based on Chl *a* or Transparency (Heiskary and Walker 1988)

Author/Location	Chl *a* (µg/L)	Rating
Walmsley (1984) South African Reservoir	0–10 10–20 20–30 >30	No Problems Scums Evident Nuisance Severe Nuisance
Barica (1975) Canadian Prairie Ponds	0–25 25–100 100–200	Clear, No Blooms Moderate Blooms Dense Colonies and Scums
McGhee (1983) North Carolina	>15 >40	Unsuitable for Trout Severe Nuisance
Lillie and Mason (1983) Wisconsin	<1 1–5 5–10 10–15 15–30 >30	Excellent Very Good Good Fair Poor Very Poor

Table 3-8. Potential Lake Uses Identified by Chl *a* Levels for Lakes in Southern Ontario (Dillon and Rigler 1975)

Chl *a* (µg/L)	Use Category
2	Body contact recreation and cold water fishery.
5	Water recreation where a cold water fishery is not imperative.
10	Body contact recreation of little importance, emphasis on warm water fish.
25	Suitable for warm water fishery.

Evidence suggests that nutrient concentrations that result in acceptable levels of algal biomass in some rivers and streams may cause unacceptable levels of algal biomass in some downstream lakes, reservoirs, and estuaries. In such cases, more stringent nutrient criteria may be required (U.S. EPA 2000a). Also, as stated previously, SSNC should be developed with a full understanding of potential impacts to downstream water quality, which could conceivably result in more strict nutrient standards than the U.S. EPA ecoregional values.

The selected Tier 1 criteria should reflect the consensus of the regulatory and regulated entities involved in proposing the SSNC. These entities should agree on the set of nutrient criteria they believe will meet the water quality objectives and designated uses established for the waterbody. The selected Tier 1 criteria should also reflect the consensus stakeholder decision. Criteria acceptable as SSNC to the stakeholders would then be submitted to the state agency and implemented as site-specific water quality standards. If the Tier 1 criteria are not acceptable, the stakeholders can use Tier 2 procedures to further develop and define the SSNC.

3.7 Implementation of Tier 1 Criteria and Nutrient Management Plans

If the proposed SSNC are acceptable to the regulatory and regulated entities involved in developing the criteria, they would be submitted to the state water quality management agency and implemented as site-specific water quality standards. Implementation would occur through the state rulemaking process for updating water quality standards, required of each state by the Clean Water Act (CWA) every three years (the “triennial review”). If these published SSNC are not acceptable to the entities, then Tier 2 or 3 procedures can be used to develop more rigorous SSNC. After the accepted SSNC are implemented as standards, nutrient management plans will be required to ensure that the standards are not exceeded. These plans may include TMDL determinations, best management practices (BMPs), discharge limitations, and other regulatory or voluntary actions.

States are required to do periodic assessments of water quality in surface waters and report those findings to U.S. EPA under Section 305(b) of the CWA. If criteria continue to be exceeded, adjustments to TMDLs or nutrient management plans would be required. Waterbodies that violate state standards are listed as impaired under Section 303(d) of the CWA, and then TMDLs must be done.

3.8 Evaluation of Tier 1 Criteria

After the SSNC and nutrient management plans are implemented, the SSNC should be evaluated to determine their success in attaining water quality objectives and designated uses. This evaluation would be accomplished using an appropriately designed monitoring plan. U.S. EPA (2000a, 2000b, and 2001) describes procedures for designing nutrient and algal monitoring plans. After the monitoring data are collected, they would be used to evaluate compliance with the SSNC and designated uses using the procedures outlined in the decision matrix presented in Section 1.3.2.

CHAPTER 4.0

Tier 2 Methods for Developing Site-specific Nutrient Criteria

4.1 Introduction

A Tier 2 analysis is initiated when Tier 1 SSNC are determined to be either underprotective or overprotective of water quality objectives and designated uses or when the Tier 1 criteria are unacceptable to regulatory agencies or to regulated entities affected by the SSNC. In Tier 2, the water quality objectives and designated uses may be reevaluated using new, additional site-specific data to determine if designated uses or other water quality objectives affected by nutrients are being achieved or maintained. The development of Tier 2 SSNC is based on models of nutrient-biological response relationships developed from data on the waterbody being evaluated. Tier 2 uses both existing data and new site-specific data. Jordan Lake, North Carolina is used as a case study example in describing the derivation of Tier 2 SSNC.

4.2 Review and Revision of Variables

Tier 1 SSNC are developed without quantifying relationships between nutrients and biological response variables. This lack of knowledge may contribute to the failure of Tier 1 analysis to produce satisfactory SSNC: efforts to reduce nutrient concentrations to achieve the Tier 1 SSNC could fail to produce the desired changes in biological response variables such as Chl *a* concentrations, water clarity, algal blooms, and DO concentrations. In addition, the selected response variables may be poorly correlated with the desired water quality objectives, including designated uses. Similarly, improper or incomplete selection of response variables could result in the failure of Tier 1 SSNC.

Tier 2 provides an opportunity to review the selected nutrients of concern and the selected response variables. This review could produce a revised list of variables more relevant to the stated water quality objectives or more closely aligned with the major impacts of nutrients on designated uses. Reducing the number of response variables may also be possible by establishing correlations among them. For example, Chl *a* and volatile suspended solids (VSS) are often correlated; measuring both could be redundant.

The primary nutrient and biological response variables will generally be the same as in Tier 1. Other influential variables, however, may be included for analysis in Tier 2. In Tier 2, more emphasis is placed on evaluating the effects of other site-specific variables that may be causing or are associated with increases or decreases in algal biomass. Tier 2 should further quantify relationships between nutrients and designated uses to support the development of more appropriate SSNC.

Variables that may confound the relationships between nutrient concentrations and algal and SAV production include volatile and nonvolatile suspended solids, turbidity, shading, transparency, flow, velocity, grazing, and substrate. For example, if a waterbody has decreased transparency because of naturally high concentrations of dissolved organic carbon or suspended sediments, photosynthesis may be inhibited and algal biomass reduced. Such relationships could justify less stringent nutrient criteria, provided downstream conditions would not be adversely affected. Secondary response variables may include DO, pH, and various measures of the structure of the fish, algal, SAV, and benthic macroinvertebrate communities. Data for the other variables that affect primary production are necessary for developing relationships between nutrients and Chl *a* concentrations, as well as for evaluating designated uses.

4.3 Augmenting Data Base

Additional site-specific primary and secondary response variable data may be required to calculate seasonal and annual trends, mean values, and variances. Several years of monthly, bimonthly, weekly, or daily data are recommended. The data must be sufficient to quantify the spatial and temporal variability of each variable in the aquatic system. If sufficient data are not available, appropriate sampling and information-gathering plans must be implemented to obtain new data. The U.S. EPA (2000a, 2000b, 2001) provides guidance on sampling plan designs for nutrient studies.

Certain site-specific response variables of interest may not be routinely collected. Such data gaps can include algal taxa and biovolume information, zooplankton data, temporal information on dissolved forms of N and P, algal toxins, particulate and dissolved organic carbon values, light penetration, chemical and physical data that describe the hypolimnetic water or chemical releases from the deep sediments, and data on macrophytes and fishes. Some information on point source discharges and nutrient loading from nonpoint sources such as urban and agricultural areas may be available from discharge monitoring reports collected by NPDES permit dischargers and submitted to state agencies and the U.S. EPA.

Relationships among nutrients, algal biomass, and water clarity often vary in waterbodies with strong longitudinal gradients (Jones and Novak 1981). Point-source nutrient additions to embayments of lakes, reservoirs, and estuaries also cause spatial differences in water quality that may preclude site-specific assessments.

Lake morphology is a key factor in determining whether internal or external processes control summertime P concentrations. The ratio of the mean depth to the square root of the surface area of a lake has been suggested as an index of the fraction of the lake's volume that determines the magnitude and the duration of summertime stratification (Osgood 1988). In shallow lakes, internal processes may regulate water quality and can decrease the benefits of external nutrient load reductions. Because of internal processes, lake restoration through nutrient reduction or biological manipulation appears to be more effective in deep lakes. Shallow waterbodies may require additional sampling effort to determine the relative importance of external and internal nutrient loading on water quality. In shallow waterbodies that mix frequently, interactions between the water and sediments can be important in determining nutrient concentrations, algal composition, and algal biomass (Bostrom et al. 1988, Marsden 1989). In these waterbodies, a large fraction of P enters the water column from the sediments during mixing events.

4.4 Data Analysis

The goals of data analysis in Tier 2 are to: 1) determine if significant relationships exist between nutrient concentrations, biological response variables, and designated uses; 2) quantify such relationships and their uncertainties; and/or 3) identify and evaluate the variables that are confounding such relationships. Tier 1 criteria estimated through the use of surrogate response variables should be revised on the basis of Tier 2 data for site-specific response variables.

Tier 2 data will broaden the analyses beyond the scope of Tier 1 by refining the quantitative link between nutrient levels and the response variables used to assess water quality. For example, the following comparisons can help define relationships between nutrients and algal biomass:

- seasonal measurements of Chl *a*, TN, and TP;
- growing season comparisons of Chl *a*, TN, and TP;
- summer mean total limiting nutrient concentration with summer mean and maximum Chl *a*;
- pre-maximum growth period (spring, pre-runoff) mean dissolved limiting nutrient concentration with maximum algal biomass;
- mean annual dissolved nutrient concentration with the 75th percentile mean algal biomass; and
- cellular concentrations of the limiting nutrient with maximum algal biomass.

If significant relationships between nutrient concentrations and Chl *a* cannot be demonstrated, possible relationships between other variables (such as suspended solids, turbidity, shading, transparency, flow, velocity, grazing, and substrate) and Chl *a* should be evaluated.

The case study example, Jordan Lake, is a reservoir located on the headwaters of the Cape Fear River between Chapel Hill and Raleigh, in the Piedmont region of North Carolina. Jordan Lake's two major tributaries are the Haw and New Hope Rivers. From Jordan Lake, the Cape Fear River flows southeasterly and enters the Atlantic Ocean at Wilmington, NC. It is classified as a "Nutrient Sensitive Water" and requires limitations of nutrient discharges. In North Carolina, water quality standards for nutrient-sensitive waters do not allow for an increase in nutrient concentrations over background levels, unless it can be shown that the increase 1) is the result of natural variations or 2) will not endanger human health, safety, or welfare and that preventing the increase would cause a serious economic hardship without equal or greater benefit to the public. North Carolina has a statewide Chl *a* standard of 40 :g/L but no TN or TP standards.

Jordan Lake is in U.S. EPA Ecoregion IX, the Southeastern Temperate Forested Plains and Hills, Level III Ecoregion 45. U.S. EPA (2000c) published data on nonreference conditions for this ecoregion. For this Level III ecoregion, 25th percentile values for TN, TP, Chl *a* and Secchi depth are shown in Table 4-1. These values could be used to set ecoregional nutrient criteria applicable to Jordan Lake.

In addition, an extensive water quality dataset for Jordan Lake was provided by Jay Sauber, North Carolina Department of Environmental Resources. The dataset included measurements from up to 20 sites throughout the lake, taken one to 12 times a year, from 1982-2000. The dataset includes a variety of nutrient measurements, Chl *a*, TSS, and Secchi depth, as well as many other variables. Table 4-1 includes Tier 1 SSNC based on 25th percentile values from Jordan Lake, which are considerably greater than U.S. EPA's ecoregional criteria.

Table 4-1. Ecoregion and Jordan Lake 25th Percentile Values for TN, TP, Chl *a*, and Secchi Depth

Source	TN (mg/L)	TP (mg/L)	Chl *a* (µg/L)	Secchi Depth (m)
U.S. EPA Ecoregion IX Lakes and Reservoirs	0.358	0.020	5.4	1.53
U.S. EPA Level III Ecoregion 45 Lakes and Reservoirs	0.304	0.023	4.9	1.66
Tier 1 SSNC for Jordan Lake, 1982–2000	0.520	0.055	15	0.45

Figure 4-1 summarizes monthly average values for TN, TP, Chl *a*, TSS, and Secchi depth for measurements made during the primary growing season, May through October, for most of the sites that had continuous monitoring from 1982-2000. Prior to about 1990, most monthly average Chl *a* measurements were greater than the state water quality standard for Chl *a*, 40 :g/L, while after 1990 most monthly average measurements were less than the standard. The drop in Chl *a* after 1990 appears to be associated with decreases in TP concentrations after 1990. TN values also decreased during this period but not as dramatically. Since about 1990, TN:TP ratios have remained about 10, which suggests that the lake is phosphorus limited. From 1982-2000, water clarity, as indicated by average Secchi depth and TSS, did not change appreciably.

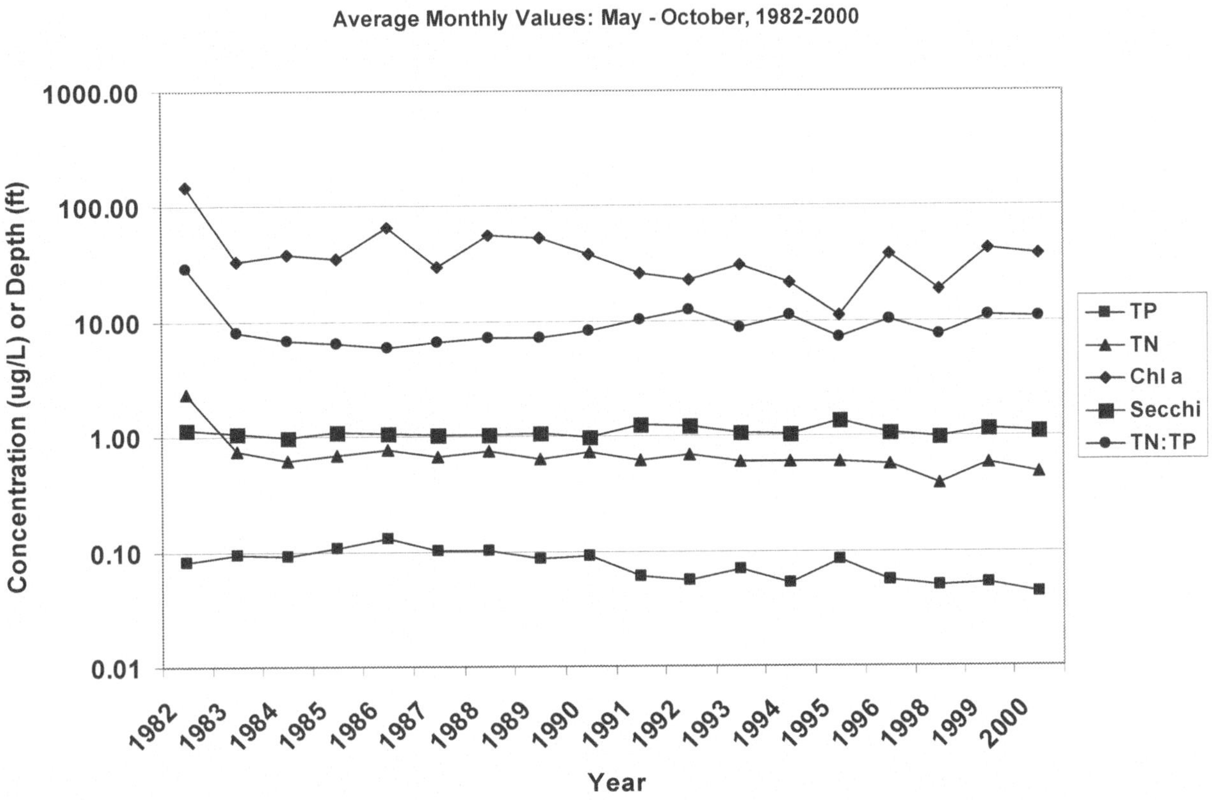

Figure 4-1. Trends in TN, TP, Chl *a*, and TSS in Jordan Lake May Through October, 1982-2000

The relationship between monthly average TN and Chl *a* during the growing season was evaluated and found to be significant (Figure 4-2), with an R^2 of 0.84.

$$\text{Chl a} = -6.6 + 65.72\text{TN} \qquad \text{(Equation 4.4.1)}$$

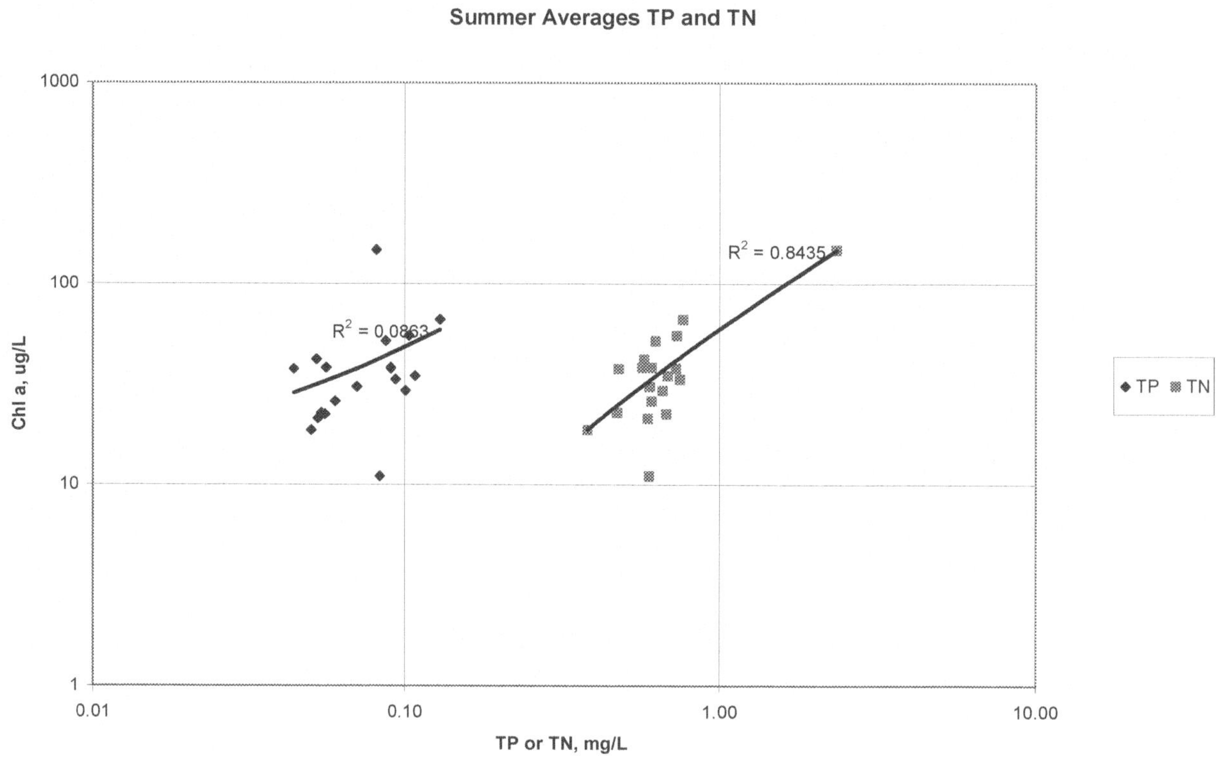

Figure 4-2. Relationship Between TN, TP, and Chl *a* in Jordan Lake May Through October, 1982-2000

The corresponding relationship for TP was not significant with $R^2 < 0.1$. Including TP in the regression model (Equation 4.4.1) did not significantly improve the R^2 value. The relationships between PO_4 and TIN and Ch *a* were also found to be insignificant.

A major confounding factor for nutrient relationships in Jordan Lake, as well as many other reservoirs, appears to be relatively high TSS levels. The high TSS causes shading, which decreases algal photosynthesis. If the source of most of the TSS in Jordan Lake is suspended mineral particles in contrast to phytoplankton, then there should not be a significant relationship between TSS and Chl *a*, which appears to be the case for Jordan Lake (Figure 4-3). It is highly likely that high mineral TSS levels are confounding any relationships between TN and TP and Chl *a* in the lake.

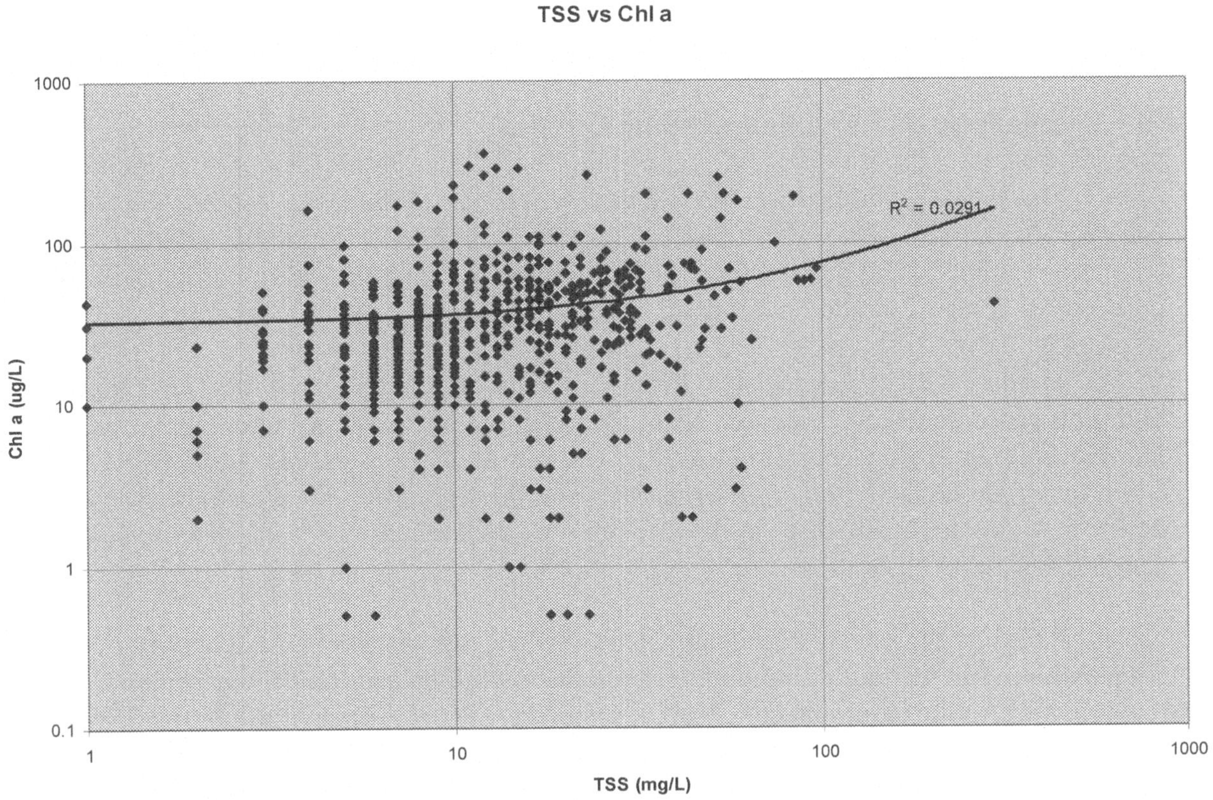

Figure 4-3. Relationship Between Chl *a* and TSS in Jordan Lake May through October, 1982-2000

4.5 Decision-making Process

The following outcomes are possible from the Tier 2 data analyses:

1. Significant relationships between nutrient concentrations and Chl *a* are demonstrated. Proceed to develop SSNC based on those relationships.
2. Significant relationships between nutrient concentrations and Chl *a* are not identified; however, the variables that confound the nutrient-Chl *a* relationships and significant relationships between these variables and Chl *a* are identified. Proceed to develop SSNC based on those relationships.
3. Significant relationships between nutrient concentrations and Chl *a* are not identified, nor are the variables that confound the nutrient-Chl *a* relationships. Proceed either to Tier 3 to collect additional data and conduct additional analyses or use the Tier 1 SSNC.

For the Jordan Lake example, a significant relationship (Equation 4.4.1) between TN and Chl *a* was found and this equation could be used to develop SSNC.

4.6 Selection of Tier 2 Site-specific Criteria

The Tier 2 SSNC are derived from the empirical relationships established between the site-specific nutrient data and the selected response variables. Using these relationships (regression models) and desired values of the response variables, interpolating the criteria for the nutrients of concern should be possible.

The first step in developing nutrient criteria from the results of the Tier 2 data analyses is to define acceptable Chl *a* biomass. Methods for doing this were discussed in Tier 1. Above these defined levels, unacceptable levels of Chl *a* is likely present. For Tier 2 SSNC, acceptable Chl *a* levels should be based on information from the waterbody for which the SSNC are being developed. As previously discussed, wide regional differences occur in what are considered acceptable Chl *a* values.

Based on the selected acceptable Chl *a* values, the empirical models developed from the data analyses can be used to predict nutrient concentrations and the 95% confidence limits that result in Chl *a* concentrations not exceeding these values. For each nutrient, the SSNC are defined as a mean value and its 95% confidence limits predicted to produce the acceptable Chl *a* values. If the predictive model uses N or P only, the Tier 1 nutrient criterion can be used as the criterion for the remaining nutrient. Alternatively, the TN: TP ratio can be used in estimating Tier 2 SSNC. As with Tier 1 criteria, less stringent SSNC should be developed only when such criteria do not adversely affect downstream waters.

The Jordan Lake example shows how SSNC can be determined. A lake user survey could be done, concluding that SSNC for Chl *a* of 28 μg/L was associated with acceptable water quality in the lake, such as water clarity, lack of algal blooms, and good fishing. This value would be 0.75 of the current standard (40 μg/L) yet about five times greater than U.S. EPA's 25th percentile ecoregional value of 6 μg/L. Based on the regression model in Figure 4-4, a monthly average growing season TN concentration of about 0.52 mg/L would have to be obtained to produce a predicted Chl *a* level of 28 μg/L. In contrast, a monthly average growing season Chl *a* concentration of 40 μg/L, the North Carolina Chl *a* standard, is associated with an average TN concentration of 0.70 mg/L.

Because of the weak relationship between monthly average Chl *a* and TP during the growing season, and the fact that Jordan Lake appears to be nitrogen limited, one method for selecting a site-specific TP criterion is to use the TN:eTP ratio of 10:1. Using this approach, if the site-specific TN criterion was set at 0.52mg/L, then the site-specific TP criterion would be 0.52/10 = 0.052 mg/L. Perhaps coincidentally, this value (0.052 mg P/L) is essentially the same as the 25th percentile TP value for Jordan Lake (0.055 mg/L) that was determined using Tier 1 methods. Table 4-2 compares the Tier 1 and Tier 2 SSNC with U.S. EPA's ecoregional criteria.

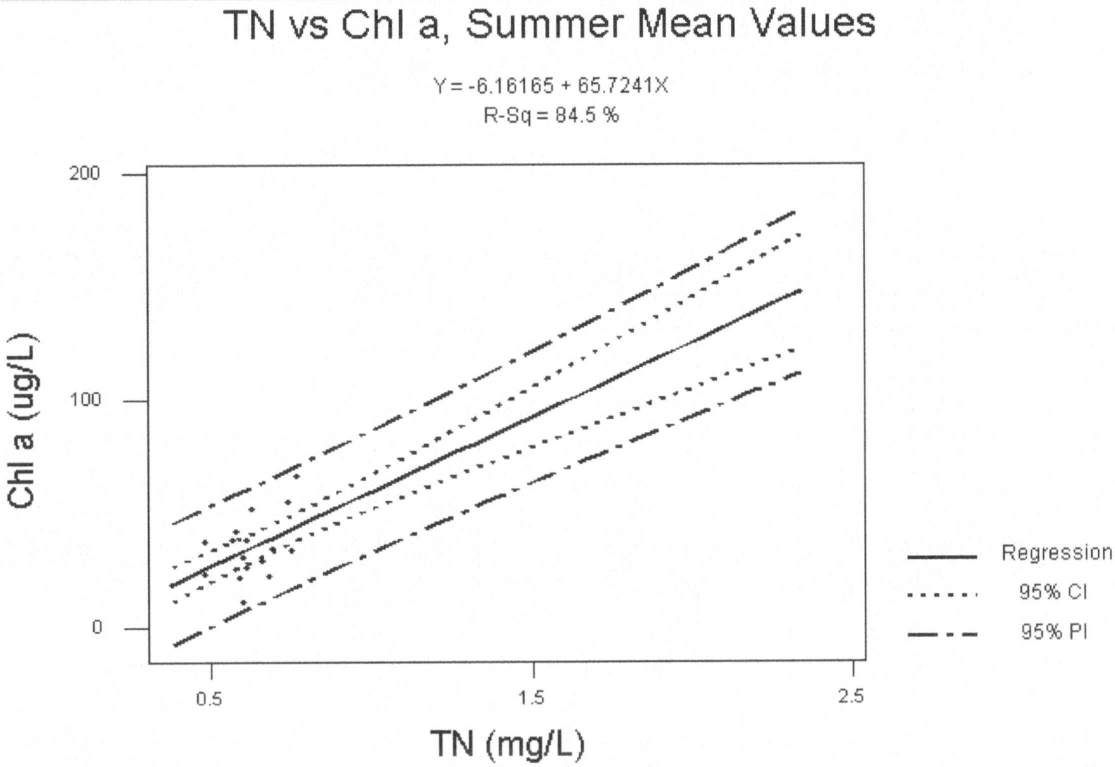

Figure 4-4. Relationship Between TN and Chl *a* in Jordan Lake May Through October, 1982-2000

Table 4-2. Tier 2 SSNC for Jordan Lake, N.C., Compared with U.S. EPA Ecoregion and Tier 1 SSNC

Source	TN (mg/L)	TP (mg/L)	Chl *a* (µg/L)	Secchi Depth (m)
Tier 2 SSNC Chl *a* = -6.2 + 66TN TN = (Chl *a* + 6.2)/66 TP = TN/10 (TN:TP = 10)	0.17 0.25 0.32 0.52	0.017 0.025 0.032 0.052	5 10 15 28	Not Determined
U.S. EPA Level III Ecoregion 45 Lakes and Reservoirs 25th Percentile Values	0.30	0.023	4.9	1.66
Tier 1 SSNC Jordan Lake, 1982–2000, 25th Percentile Values	0.52	0.055	15	0.45

4.7 Implementation of Tier 2 Criteria and Nutrient Management Plans

If the proposed SSNC are acceptable to stakeholders, the SSNC would be submitted to the state water quality management agency and implemented as site-specific water quality standards. Implementation would occur through the state rulemaking process for updating water quality standards, required of each state every three years by the Clean Water Act (the "triennial review"). If these published values are unacceptable to stakeholders, Tier 3 procedures can be used to develop alternative SSNC.

After the SSNC are implemented as standards, nutrient monitoring and management plans will be required to ensure that the standards are not exceeded. These plans may include TMDL determinations, BMPs, discharge limits, and other regulatory or voluntary actions.

The Tier 2 models can be used to determine the potential benefits of nutrient reduction programs resulting from land management efforts or advanced wastewater treatment. Nutrient reductions should have direct impacts on algal levels, oxygen content, and water clarity. The magnitude of these changes can be estimated with the empirical models. As an example, Jordan Lake is known to support algal blooms during the summer, which decrease water clarity and reduce fishing success. A substantial source of nitrogen load to the lake is the discharge from a permitted point source located within its watershed. Assuming that a nitrogen load evaluation was done, attributing half the annual nitrogen load to this facility, and that advanced wastewater treatment will reduce average nitrogen values in the lake from 0.70 mg/L to 0.52 mg/L, the empirical model in Figure 4-4 indicates that reductions in TN of this magnitude should reduce average summer Chl *a* values from ~ 40 µg/L to ~ 28 µg/L, which is the SSNC.

4.8 Evaluation of Tier 2 Criteria

After the SSNC and nutrient management plans are implemented, they should be evaluated to determine their success in attaining water quality objectives and designated uses. This evaluation would be accomplished using an appropriately designed monitoring plan. U.S. EPA (2000a, 2000b, and 2001) describes procedures for designing nutrient and algal monitoring plans. After the monitoring data are collected, they would be used to evaluate compliance with the SSNC and designated uses using the procedures described in Section 1.4.6.

CHAPTER 5.0

TIER 3 METHODS FOR DEVELOPING SITE-SPECIFIC NUTRIENT CRITERIA

5.1 Introduction

A Tier 3 analysis can be implemented if the Tier 2 process does not provide SSNC that achieve the desired water quality objectives. Tier 3 methods and approaches may be used to revise SSNC that have been demonstrated as inadequate (not protective or too stringent) as a result of Tier 2 monitoring and evaluation programs. Tier 3 may also be undertaken if a more comprehensive evaluation of an existing nutrient standard is warranted.

The Tier 3 assessment process parallels Tier 2 with two important modifications (see Chapter 1, Figure 1-7): 1) Tier 3 emphasizes the collection of site-specific data necessary to support the derivation of SSNC, and 2) Tier 3 analyses focus on process-level evaluation of the site-specific relationships between nutrients and the values of the selected response variables.

Two objectives of the Tier 3 assessment are to: 1) produce SSNC, if such criteria were not produced by either Tier 1 or Tier 2 analyses; and 2) modify or refine criteria produced by the Tier 2 assessment. The desired results of the Tier 3 assessment will be SSNC that have been developed by collecting as much new data as time and funding resources permit and applying state-of-the-art, quantitative scientific understanding of nutrient dynamics. Alternatively, the Tier 3 assessment may conclude that, given the newly available data and current understanding of the aquatic systems of interest, SSNC are not feasible.

The Tier 3 assessment continues the Tier 2 emphasis on constructing a process-level understanding among nutrient loadings, nutrients concentrations, and the corresponding magnitude of response by the ecological resources relevant to the stated water quality objectives. Tier 3 may extend the statistical models implemented as part of Tier 2 by incorporating mathematical and computer simulation models that have been developed to characterize the relationships among nutrient loading, subsequent concentrations of nutrients, and the production dynamics of critical populations of estuarine organisms.

5.2 Review and Revision of Nutrient Response Variables

Tier 3 provides a final opportunity to review and revise the response variables selected for developing the SSNC. In practice, the set of variables expanded in Tier 2 probably will not change dramatically in Tier 3. The primary nutrients and biological response variables carried through Tiers 1 and 2 will likely remain the focal points of the assessment. At the same time, variables that proved uninformative in Tiers 1 and 2 may be omitted or replaced in Tier 3.

Response variables may also be added in Tier 3. Tier 3 attempts to develop a more mechanistic understanding of the physical, chemical, and biological processes that determine the values of selected response variables in relation to site-specific nutrient loading. As will be described, complex ecological models of aquatic systems may be used as the operational framework for developing this understanding and specifying SSNC. As a result of using a specific model, additional response variables (model state variables) not previously considered in Tier 1 or 2 may necessarily enter into the assessment. This possible means of identifying additional response variables sets Tier 3 apart from Tiers 1 and 2, which are primarily data driven.

5.3 Augmenting Data Base

More extensive data than used for Tier 2 analyses will be required for the aquatic system of interest. If Tier 3 analyses focus on the use of *in situ* experiments to develop a mechanistic basis for generating SSNC, additional resources may be required in order to carry out site-specific experimental designs. In addition to quantifying the spatial and temporal variability of previously examined Tier 2 response variables, the site-specific implementation of an aquatic ecosystem model in Tier 3 may well require the collection of additional data needed for the model (incident surface light intensity, water temperature, seasonal discharge data, bathymetry of the segment, measures of benthic invertebrates and fish).

5.4 Data Analysis and Process-level Modeling

The empirical Tier 2 analyses relating Chl *a* to total N or P often shows a high degree of variability in these relationships. The imprecision in these statistical regressions may prove unacceptable to stakeholders for deriving the SSNC. As a result, the stakeholders may choose to use a more complex, process-level model and attempt to establish a more accurate and reliable basis for deriving site-specific criteria.

Table 5-1 lists several U.S. EPA models that may be applied on a site-specific basis to examine relationships between nutrient loading, bioavailability of nutrients, and subsequent values of algal biomass (Chl *a*), as well as other response variables of interest. Other published models in the research and development literature may also be relevant for site-specific criteria, for example, the Integrated Fates and Effects Model (IFEM) (Bartell et al. 1988) and the Comprehensive Aquatic System Model (CASM) (Bartell et al. 2000). In practice, a comprehensive search of the technical literature and available government resources should be performed to identify the model most appropriate to the aquatic system of interest.

Table 5-1. Models for Deriving Site-Specific Nutrient Criteria

HSPF	Hydrological Simulation Program-Fortran
SWMM	Storm Water Management Model
WASP	Water Quality Analysis Simulation Program
BASINS	Better Assessment Science Integrating Point and Nonpoint Sources
QUAL2E	Enhanced Stream Water quality Model
IFEM	Integrated Fates and Effects Model
CASM	Comprehensive Aquatic System Model for risk assessment
AQUATOX	Aquatic ecosystem model for assessing risks from chemicals

5.4.1 Model Selection, Guidance, and Calibration

Interactive use of models, field studies, and controlled experiments can contribute the understanding of complex ecological systems (Madden and Kemp 1996). A credible modeling framework can be used to examine cause and effect relationships between nutrient enrichment and biological responses relevant to site-specific designated uses. Models can provide quantitative estimates of system response to nutrient perturbations and management.

Importantly, although the development and application of models to the management of nutrients continues to evolve, areas of improvement remain. For example, Thomann and Linker (1998) identified three issues in model development and application to nutrient control and management of eutrophication:

- degree of controllability of nonpoint sources of nutrients;
- interaction of hydrodynamic transport, sediments fluxes, and water quality; and
- significance of benthic suspension feeders in influencing water quality.

These challenges to model development have emerged particularly in the assessment of nutrients in estuarine (the Chesapeake Bay) and coastal waters, although analogous considerations may well apply to models of nutrient dynamics in streams, rivers, lakes, and reservoirs. The following sections address issues relevant to the use of ecological and environmental (hydraulic, physical transport) models in deriving site-specific nutrient and other water quality criteria but do not provide a comprehensive, in-depth discussion of ecological and environmental modeling. Such reference texts on ecological modeling are available (Patten 1971, Odum 1983, Swartzman and Kaluzny 1987). The intent here is to outline some of the technical issues that should be addressed in selecting and using models for deriving SSNC.

The discussion mentions by way of example several water quality and eutrophication models that have proven useful for quantifying nutrient dynamics in aquatic systems. However, reference to any specific model is not intended as an endorsement or recommendation of the model for estimating SSNC.

5.4.1.1 Modeling Objectives

Of major importance in selecting models for estimating SSNC are the water quality and modeling objectives designated for the site of interest. Models should be selected that directly address the SSNC objectives defined by managers and stakeholders. For example, if site-specific designated uses emphasize biocriteria, models should be selected that translate nutrient loading or nutrient concentrations into estimates of nutrient concentrations as well as the biomass of phytoplankton, periphyton, rooted aquatic plants, fish, and other organisms of concern. Given the likely number and diversity of management objectives relevant to SSNC, more than one model may be required to comprehensively address the nutrient management objectives.

In guiding the selection of models for evaluating SSNC, water quality managers must recognize that models are used differently in environmental management compared to basic research and development (Campbell and Bartell 1998). The objective in basic research is to falsify the model and thereby determine the necessary refinements and modifications consistent with advancing science and subsequently improving model performance. In contrast, managers desire models with known accuracy and reliability to use with confidence in environmental decision-making. Models that remain under development by the basic research community may nonetheless be applied in estimating nutrient criteria, if the performance of the model has proven sufficient for known combinations of parameter values, initial conditions, and forcing functions that map onto aquatic systems of interest.

5.4.1.2 Model Structure

In selecting among existing models, adapting models, or developing new models for deriving SSNC, water quality manager need to evaluate aspects of model structure. These structural issues include model complexity, scale, governing equations, process-level formulations, and external forcing functions. Parameter estimation is not considered a component of model structure and will be addressed as a separate category in relation to model selection and implementation for SSNC.

5.4.1.2.1 Complexity

Model structural complexity, also termed *articulation*, refers to the degree of structural and process-level detail included in a simulation model. In a “box-and-arrow” schematic illustration of a simulation model, the number of boxes (state variables), as well as the number and arrangement of arrows (functional interconnections of the state variables) characterize the degree of model complexity. In mathematical terms, the number of coupled equations that constitute the model defines structural complexity. Model complexity may also be interpreted as the number of independent variables in a statistical (multiple regression) model.

In some applications, simpler models may prove as informative as more complex models designed to address similar objectives. For example, Hagy et al. (2000) demonstrated the effectiveness of using a “box-model” approach to accurately estimate hydraulic residence times for the Patuxent River estuary, Maryland. This comparatively simple model structure was offered as an economic substitute for much more complex two- or three-dimensional hydraulic models (Cerco and Cole 1993). The box model provided useful information describing transport and residence times necessary to understand the potential impacts of nutrient loading on water quality and biological production dynamics in the Patuxent River and estuary.

There is no general theory to guide the degree of complexity necessary in an ecological model, although a considerable understanding of the implications of aggregating model structures into simpler models (aggregation error) has been developed (Bartell et al. 1988, O'Neill and Rust 1979, Cale et al. 1983, Zeigler 1976). Nevertheless, some general guidance can be offered. Model developers should strive to create models that are of sufficient structural and functional complexity to provide as accurate as possible descriptions of the ecological phenomena of interest. Additional complexity that fails to increase the accuracy of the model results should not be retained as part of the model. Simple models are sometimes recommended from the viewpoint of model transparency, or ease in understanding the model, particularly among nonmodelers. Overly simplistic models, however, may fail to provide results of sufficient accuracy or reliability to be useful in deriving SSNC.

5.4.1.2.2 Scale

Models are designed to describe nutrient dynamics for some specified (or implicit) spatial and temporal dimension. The spatial scale of a model includes *grain* and *extent* (Allen and Starr 1982). Grain defines the smallest spatial or temporal resolution of the model; extent delineates the overall spatial dimension or time period addressed by the model. For example, a water quality model may simulate nutrient dynamics for a square kilometer of surface area and corresponding bathymetry (extent) by dividing the system into a set of one-cubic meter spatial elements (grain). The time-step of a model (minutes for some complex hydrodynamic models) would define the grain in the time domain, while the simulated time period (one year, for example) would delineate the analogous temporal extent.

Model selection must include consideration of the appropriate scales (in time and space) relevant to the ecological and water quality issues addressed by the model (Fitz et al. 1996). Scaling issues are important in understanding and modeling nutrient dynamics. For example, Gallegos et al. (1992) modeled the local nutrient inputs from watersheds adjacent to the Rhode River-Muddy Creek subestuary and inputs from more remote watersheds. The model indicated that a single rainfall event resulted in nutrient loadings to the mainstem estuary at two scales. Direct loading from local sources produced short-term, smaller scale phytoplankton blooms, while loadings from more remote watersheds were associated with larger scale, longer term blooms. These results were corroborated by observations made for two separate rainfall events.

Considerations of scale also apply to model selection in relation to the scales of resource management and decision making (Peterson and Parker 1998). Care must be taken to avoid the selection of models that are disparately scaled from the spatial-temporal framework relevant to managing or controlling nutrient dynamics in the derivation of SSNC.

5.4.1.2.3 Governing Equations and Process Formulations

The previous discussion of model structure addresses what is included in the model. Model equations and process formulations define how the structural components of a model interact to accomplish the model objectives and how model input values translate to model results. Again, a comprehensive discussion of the nature of governing equations and process formulations used in models relevant to SSNC exceeds the scope of this report. Nevertheless, some brief description of commonly encountered mathematical equations and formulations may assist in an educated selection or development of models for SSNC.

In data-based and derived statistical models (Tier 2), the basic model form will likely be a linear or nonlinear regression equation that estimates model results as a function of regression coefficients applied to the model's independent variables. Individual regression terms (operationally analogous to process formulations in simulation models) range from linear, first-order coefficients to higher order, nonlinear terms. Models that estimate chlorophyll concentrations from total annual phosphorus loads (example Tier 2 models) demonstrate these kinds of equations. These models can be sufficiently accurate in their predictions and are often more easily implemented than complex simulation models. One limitation of this model form lies in the difficulty in interpreting the regression coefficients or individual model terms in the context of physical, hydrodynamic, chemical, or ecological processes that can be controlled or managed. Another potential constraint in using empirical models for SSNC is the extrapolation of the model for values of independent variables that are outside the domain of data used to develop the model.

Governing equations and process formulations of simulation models (loading models) attempt to describe in mathematical form the physical, chemical, and biological mechanisms that determine changes in the values of the model's structural components (state variables). Systems of coupled differential (or partial differential) equations have been routinely used to describe the dynamics of populations of aquatic organisms in eutrophication models relevant to SSNC (Di Toro et al. 1971). These equations are often linear and first-order, although nonlinear formulations of mechanistic processes (nutrient limitation of photosynthesis, grazing, predation, decomposition) are also commonly incorporated into these systems of governing equations (Bartell et al. 1999, DeAngelis et al. 1989). Because the mechanisms underlying the dynamics of the modeled system are thought to apply universally, the process-level models are designed to be broadly applicable given site-specific values of model parameters, initial conditions, and external forcing functions (nutrient loading, water temperature, surface light intensity). The main challenges in using these kinds of models in deriving SSNC include the resources needed to estimate model parameters for the system of interest and the degree of accuracy and reliability of model results. The advantage of this approach lies in the attempt to describe system dynamics using real-world measurable processes; correspondingly, the process formulations and parameter values can be readily understood in the context of nutrient management plans and actions. Nonetheless, the level of expertise and effort required in using these kinds of models for SSNC suggests that they be reserved for Tier 3 assessments of SSNC.

5.4.1.2.4 External Forcing Functions

The more complex simulation models characteristic of Tier 3 SSNC derivation commonly require external data to perform model calculations. Such data may take the form of daily nutrient loading rates, water temperature, water current velocity, water elevation, and solar radiation. In waterbodies such as reservoirs or estuaries, complex hydrodynamic models may have to be implemented in order to provide the necessary inputs to water quality models used to estimate SSNC or to evaluate the implications of management scenarios on water quality or attainment of designated uses. Selection of models may be influenced by the availability of required site-specific data inputs or the amount of resources necessary to obtain such information.

5.4.1.3 Model Parameter Values

Accurate and precise values of model parameters are essential when using models to estimate SSNC or evaluate management alternatives directed at nutrient control. Consideration of candidate models should include two key issues: 1) the feasibility of providing site-specific values for the necessary parameters, and 2) the ability to characterize parameter uncertainty and examine the implications of these uncertainties on model results.

5.4.1.3.1 Estimation

Two criteria for selecting models for use in SSNC include the feasibility of obtaining accurate and precise estimates of parameter values and the amount of resources (sampling effort, time, manpower, funds) required to obtain these values. These criteria pertain mainly to process or simulation models, where the parameter values have operational and measurable meaning in the context of physics, hydrology, ecology, chemistry, and ecology. However, empirical or statistical models can be implemented for new sites by collecting the necessary data describing the dependent and independent variables in the model, and then rederiving the model (estimating the statistical parameters) through the appropriate statistical procedures (regression). Regardless of the model formulation, key criteria in selecting among models for use in nutrient management or defining SSNC focus on the feasibility and effort necessary for parameter estimation. Clearly, the greater the complexity of the model, the greater will be the challenges and resource demands in obtaining accurate and precise parameter values.

5.4.1.3.2 Uncertainty

The environmental processes represented in simulation models or summarized statistically in empirical models are inherently variable or uncertain. Supporting parameter values are similarly uncertain. A comprehensive discussion of the nature and sources of uncertainty in environmental models is not intended in this discussion (see Warren-Hicks et al. 1998). Simply, bias and imprecision can enter into parameter estimation as a result of natural variability, sampling error, inappropriate measuring methods, and incomplete or incorrect understanding of system structure and function. Depending on the nature of the model, uncertainties in parameter values may be characterized in the form of intervals, fuzzy numbers, or statistical distributions. Models that permit the characterization of parameter uncertainty may be selected over models that lack this ability.

5.4.1.4 Model Evaluation

Perhaps the most important criteria, apart from the ensuring that a candidate model addresses one or more of the SSNC modeling objectives, include evaluation of model performance. Again, in the context of nutrient management and control, the model user wants to know or anticipate the degree of accuracy and reliability of model results in site-specific applications. There are several aspects to consider under the broader category of model evaluation: 1) calibration, 2) model: data comparisons, 3) perturbation experiments, and 4) sensitivity and uncertainty analyses.

5.4.1.4.1 Calibration

Calibration refers to the process of adjusting model parameter values to maximize the correspondence between model results and observations (data). A variety of methods exist for quantitatively characterizing the "goodness of fit" between model calculations and observations (Bartell et al. 1992). Clearly, the nature and extent of model calibration will be determined by the availability of data. Data needs for calibration are determined by the complexity and scale of the model. Thus, benchmark calibration data sets for complex ecosystem models are resource intensive and are sometimes compiled using data collected from several similar systems or the same system at different times (Madden and Kemp 1996). The value of comprehensive and quality data sets for calibration and model: data comparisons (see below) cannot be overestimated. Importantly, data used in deriving model process formulations or parameter values should not be used in calibration.

By identifying critical model parameters, calibration can be guided by the results of sensitivity or uncertainty analyses of the model. Ideally, calibration parameters should correspond to physical, chemical, or biological processes, and be estimable from field data, controlled experiments, or other models. For effective use in nutrient management, the key model parameters ought to refer to components of the system related to nutrient dynamics that can be managed and that respond to management actions.

5.4.1.4.2 Model: Data Comparison

Perhaps the most compelling evaluation of model performance is the comparison of results from a previously calibrated model to independent sets of field data. As the number of a model's structural components increases, the challenges in obtaining simultaneous accurate predictions of all components increase correspondingly. (This truism has been sometimes used illogically to argue for simpler models.) Repeated comparisons of model results with observations obtained from different systems can be used to identify the kinds of systems that the model usefully represents, as well as those systems for which the model does not provide accurate predictions. These model "successes" and "failures" provide valuable information for selecting models for SSNC estimation.

The required accuracy in model performance for the estimating SSNC will likely be determined by water quality managers in the decision-making process. Stakeholders will undoubtedly stipulate the desired model performance objectives. Model performance criteria in a decision-making context may be specified in relation to the consequences of incorrect model projections.

5.4.1.4.3 Perturbation Experiments

Model calculations and data may correlate for the wrong reasons. That is, the model processes leading to the goodness of fit are not those ecological and environmental processes actually producing the data. Perturbation experiments can address this potential problem in model evaluation. Perturbation experiments require manipulation (nutrient enrichment) of the ecosystem and performing the corresponding manipulation to the model (simulated nutrient enrichment). If the model accurately describes the observed responses of the measured system to perturbation, additional credibility may be conferred on the ability of the model to provide accurate and reliable results relevant to derivation of Tier 3 SSNC.

5.4.1.4.4 Sensitivity and Uncertainty Analysis

Strictly defined, *sensitivity analysis* determines the partial differentials of model outputs to model input values (Tomovic and Karplus1963). For models amenable to analytical calculus, the partial differentials can be formulated and solved explicitly. For most models, however, sensitivities are numerically approximated (using Monte Carlo simulation) by characterizing the impacts on model calculations of small variations in the input parameter values (1% of the nominal values, Gardner et al. 1981). This analysis emphasizes the impacts on model calculations per unit uncertainty in model parameters. The key parameters identified in a sensitivity analysis provide information on the mathematical structure of a model; for example, exponents in model formulations tend to be important structural parameters per unit variability. The results of sensitivity analysis are particularly useful if no information is available concerning the likely uncertainties associated with model parameters. Sensitivity analysis is also useful in the process of model development, prior to the estimation of model parameters. Parameters that are potentially important simply as the result of the mathematical structure of the model (exponents) can be readily identified and efforts can be correspondingly directed towards collecting the necessary data for accurate and precise estimation of sensitive parameters. Alternatively, restructuring or reformulation of the model can be considered in order to reduce sensitivities owing simply to model structure.

In contrast, *uncertainty analysis* (sometimes referred to as *error analysis*) examines the response of model results to variations in parameter values that reflect the uncertainty expected or measured from field data. Instead of focusing on model structure, an uncertainty analysis characterizes the uncertainties associated with model results in practical applications of the model. In a numerical approach to uncertainty analysis (Monte Carlo simulation), input parameters are characteristically assigned different distributions that best describe the variability associated with each particular parameter in site-specific applications. Key parameters in an uncertainty analysis identify those model inputs that contribute most to variability in model performance, given the kinds of bias and imprecision estimated for the site-specific application of the model for SSNC derivation.

As an example, Fong et al. (1997) demonstrated the utility of uncertainty analysis of a seagrass model in understanding the complex relationships between nutrient impacts on seagrass production. The analysis indicated that under low-nutrient conditions, factors including light, temperature, and salinity controlled seagrass production. When nutrients were abundant, epiphyte production and subsequent shading of seagrass leaves were the dominant factors influencing seagrass production.

Models that are amenable to sensitivity and uncertainty analyses or models that have been analyzed in this manner are recommended for defining SSNC over models that make difficult such analyses.

5.4.1.4.5 Other Evaluation Criteria

The preceding discussion emphasized the technical aspects of model evaluation. However, the model selection process may also consider if the model has been used previously for similar objectives relevant to nutrient dynamics. Additional consideration may be given to the general acceptance of the model by the scientific community. Finally, the degree of model credibility among regulators and decision makers can be a factor in the comparison and evaluation of alternative models for SSNC. Although these criteria may prove more qualitative and subjective in application, these considerations can in some cases determine the selection of models (Pastorok et al. 2002).

5.4.1.5 Model Documentation

The selection of models for site-specific applications can be facilitated if documentation is available that adequately describes the model of interest. Such documentation should describe the overall objectives for developing the model. The science underlying the model structure should be clearly discussed. The structure of the model should be clearly presented. Empirical relationships or process formulations should be presented in detail. Sources and previously used values of model parameters should be provided; values should be justified or substantiated. Example calibrations and model performance in prior applications should be provided. The results of any previous sensitivity or uncertainty analyses should be included in the documentation. Model selection will probably require special expertise and guidance that may be provided by a consultant or technical advisory committee.

Documentation is also a necessary aspect in using selected model(s) to estimate SSNC. Documentation should include detailed descriptions of the assumptions used in model implementation, descriptions of any modifications to the model, selection and justification of parameter values, and description and interpretation of model results.

5.4.2 Recommendations for Model Selection

In selecting a model to use in deriving SSNC, the following recommendations are offered:

1. Thorough consideration and careful thought should be given to the general and specific objectives of the modeling effort. The modeling objectives should be stated with as much precision as possible, within the overall nutrient management and decision-making process.

2. The modeling objectives should guide the initial selection of potentially applicable models. All the modeling objectives may not be addressed by a single model. Thus, a set of models may be required.

3. If models are not available for the specific system of interest, models should be chosen that have been constructed for similar systems. A hierarchical approach to evaluating similarity may reasonably address geomorphology (or bathymetry), land use, hydrology, climate (meteorology), nutrients, producers, and consumers (Paul 2001).

4. The complexity of the selected models should be commensurate with the requirements of the modeling objectives. However, a parsimonious approach to model complexity is suggested, where simpler models are selected over increasingly complex models, if the models demonstrate similar performance in relation to the objectives.

5. It may prove effective to progress from simple empirical (regression) models to highly aggregated ecological models and finally to complex, linked hydrodynamic and ecological models in selecting and implementing models for developing SSNC.

6. The characteristic scales (time and space) of structures and processes that determine site-specific nutrient dynamics for the system of interest should be represented in the selected models. Inconsistencies in scale between nutrient management objectives and model capabilities should be avoided.

7. The feasibility of estimating site-specific values for model parameters should influence the selection of models for nutrient criteria. Key model input values should be readily measurable and verifiable for the system of interest.

8. The availability of external data requirements, in addition to model parameter estimation, may determine the feasibility of implementing a candidate model for SSNC. Such requirements can guide the selection among alternative models.

9. The performance characteristics ("goodness of fit") for selected models should be known for applications to the system of interest or for similar systems for which the models have been previously applied.

10. The selected model(s) should be amenable to quantitative sensitivity and uncertainty analysis. The results of these analyses can be used to guide model refinements as necessary to improve performance in estimating SSNC.

5.5 Examples of Tier 3 Methodology

The following sections describe example derivations of Tier 3 SSNC. To provide as broad a perspective as possible, these examples are hypothetical—models were not implemented, and calculations were not performed.

5.5.1 Rivers and Streams

For the Little Maries River example (see Section 5.4), a decision to proceed to a Tier 3 analysis was based on the desire to develop more accurate and precise relationships between TN, TP, and Chl *a*. After a detailed evaluation of potentially useful models, a decision was made to implement a site-specific version of the QUAL2E model (Brown and Barnwell 1987). The necessary data for applying the model were developed from the Tier 3 data augmented by the collection of additional information for the Little Maries River. Using the model, site-specific data, and different assumptions concerning N and P loading rates from the smaller tributaries, sets of simulations could be performed to examine the implications of nutrient loading on Chl *a* values.

Simulated loading rates for QUAL2E could be determined consistent with the measured values of TN and TP concentrations corresponding to the Tier 2 Chl *a* values that ranged between 50 and 100 mg/m^2. For each nutrient loading scenario, the QUAL2E model could produce an annual series of daily values of dissolved nutrients (N, P) and corresponding values of Chl *a*. The results of repeated simulations could produce sets of model output that provide the same kinds of descriptions developed for Tier 2, namely quantitative relationships between concentrations of TN, TP, and algal chlorophyll *a*. In Tier 3, however, these relationships reflect the basic physical, chemical, and biological processes formulated in the QUAL2E, rather than the strictly empirical regressions used in Tier 2.

For the Little Maries River case study, the Tier 2 regression models suggested that the 100 mg/m^2 value of Chl *a* was consistent with nutrient criteria of approximately 2,120 :g/L total N and 911 :g/L total P. The more protective Tier 2 criteria to obtain a 50 mg/m^2 value for Chl *a* were 474 :g/L for total N and 51 :g/L for total P. Hypothetical results of repeated simulations involving different assumptions concerning nutrient loading and the conversion of nutrients to algal biomass (modeled algal growth parameters) could indicate that site-specific criteria for N and P consistent with a desired Chl *a* value of 100 mg/m^2 are lognormally distributed with a geometric mean value of 1,839 :g/L for total N (geo. st. dev 371 :g/L) and a geometric mean value of 723 :g/L of total P (geo. st.dev 238 :g/L). These kinds of Tier 3 modeling results suggest that the empirically Tier 2-derived criteria were overprotective.

5.5.2 Lakes and Reservoirs

Table 4-2 summarizes the Tier 1 and Tier 2 SSNC developed for Jordan Lake, North Carolina and compares them with the U.S. EPA regional criteria. These criteria were developed on the basis of understanding relationships between concentrations of nutrients and Chl *a*. Tier 1 also addressed water transparency (Secchi depth). A Tier 3 analysis for Jordan Lake could be stimulated by the failure of Tier 2 SSNC in achieving and maintaining designated uses or by a desire among stakeholders to incorporate additional ecologically defined water quality characteristics, possibly including the production dynamics and population sizes of valued recreational fishes (largemouth bass) and their forage base (aquatic insects, other invertebrates, smaller fishes). Following such a decision, the challenge becomes one of developing quantitative relationships between nutrient dynamics and the spatial-temporal patterns of abundance of these key components of the Jordan Lake food web.

If there are sufficient data that describe the varying abundance of the aquatic populations of interest, the Tier 3 analysis may closely resemble the empirical approach summarized in key regression models that resulted from the Tier 2 assessment (Figure 4-4). Several simple regression models be developed that, for example, relate nutrients to chlorophyll, chlorophyll to zooplankton, and zooplankton to forage fish biomass. Using this sequence of empirical analyses, it would be possible to determine nutrient concentrations consistent with a specified, desired abundance of forage fish. Alternatively, multivariate models could be developed that simultaneously examine all the available data and provides a single statistical model (linear or nonlinear multiple regression) that directly estimates forage fish biomass from the selected set of independent variables. Again, building this understanding of indirect effects of nutrients on food web production dynamics using an empirical approach could require amounts of data that are seldom available. New data would have to be collected to implement this statistical approach to the Tier 3 derivation of SSNC for Jordan Lake.

Alternatively, the Tier 3 analysis could be based on a water quality model (QUAL2E) or aquatic ecosystem model (AQUATOX, CASM) appropriate for lakes and reservoirs. The criteria for selecting among available models are discussed in Section 5.4.2.

Given an appropriate model, the necessary site-specific input data would have to be developed for Jordan Lake. Realistically, the model application may require the use of data for other similar reservoirs in combination with existing data or new data, depending on the available resources that could be directed at the collection of new data for Jordan Lake. If possible, the model inputs should include quantification of spatial and temporal variability, as well as bias and imprecision (uncertainty). The implications of such uncertainties on the resulting derivation of Tier 3 SSNC could be examined using repeated simulations with values of model inputs selected in a systematic (interval analysis) or statistical (Monte Carlo) fashion. Some models have been developed in a Monte Carlo framework (the CASM) to facilitate these kinds of analyses.

The results of the Tier 3 aquatic system model analysis could take the form of average annual biomass of selected aquatic populations of interest. Alternatively, time series (daily values) of population biomass and nutrient concentrations could be produced by the model. If uncertainties are included in the analysis, these model results could be described by distributions of average annual biomass or multiple time series. These kinds of results could be used to identify nutrient criteria consistent with desired population sizes of the aquatic food web components of interest to stakeholders. Depending on the choice of model, concentrations of dissolved oxygen, as well as TSS and phytoplankton biomass (water transparency) could be estimated for different nutrient loading scenarios. These results could also be used in the derivation of Tier 3 criteria. The Tier 3 criteria generated in this manner for Jordan Lake could then be compared with the results of the Tier 1 and Tier 2 values.

5.5.3 Estuaries

Numerical models of coupled biological and hydrodynamic processes in shallow tidal systems (Lucas et al. 1998) could be implemented to provide more site-specific, process-oriented descriptions of the relationships between nutrient loads and biological responses (phytoplankton blooms) for a Tier 3 Neuse River estuary analysis, for example. Similar modeling approaches to describe dinoflagellate blooms and microbial food web production dynamics (Eldridge and Sieracki 1993) may be usefully applied in Tier 3 assessments. Table 5-2 lists other estuarine models that could be used for derivation of Tier 3 SSNC. The listed models are not intended to comprehensively represent all available models that could prove useful in developing SSNC for estuaries. Nor does listing in this table imply any particular endorsement of a model for SSNC. Rather, the listed models include those that meet at least one or more of the recommended criteria for selecting models to be used in arriving at Tier 3 SSNC. The use of any of these models is likely to be costly due to the amount of data needed to calibrate the models, time needed to run the models, and expertise needed to interpret the model outputs.

Given the complexities in model selection and resources required for model implementation, the use of process models for SSNC described below is strictly hypothetical.

Figure 5-1 shows hypothetical time-varying values of primary production produced by a dynamic estuarine ecosystem model. Such time series can be analyzed to develop SSNC, specified either as a criterion maximum concentration (CMC), or criterion continuous concentration (CCC), or an external loading rate to the system. In developing or implementing existing process-level simulation models for SSNC, the nature of the decision-making process and management alternatives should be used to identify the necessary and sufficient degree of detail to be included in the models.

Table 5-2. List of Example Models Potentially Relevant to Tier 3 Development of SSNC

Model	Description	Reference
Cumberland Basin model, BOEDE EMS	Holistic model describing production dynamics in the Cumberland Basin estuary	Gordon et al. (1986)
GEMBASE	General ecosystem model for the Bristol Channel and Severn Estuary, UK	Radford (1979)
Delaware Estuary	Model of nutrients, algae, zooplankton, fish and decomposers, biological oxygen demand	Kelly and Spofford (1977)
Narragansett Bay	Integrated physical, chemical, and food web model for nutrients and energy flows	Kremer and Nixon (1978)
Barrataria Bay	Model of carbon and nitrogen flows in open estuarine system	Hopkinson and Day (1977)
Estuarine flow analysis model	Food web analysis for estuarine systems	Deegan et al. (1994)
3-D eutrophication model of Chesapeake Bay	Spatially explicit hydrodynamic model of bay and tributaries, water column and sediment nutrient dynamics	Cerco and Cole (1993)
Submersed plant community model	Ecological production dynamics of submersed aquatic plants in relation to light, temperature, nutrients	Madden and Kemp (1996)
GEM	General ecosystem model developed for the Patuxent River estuary	Fitz et al. (1996)
Hydraulic residence model	Simple box-model approach for simulating water residence times in coastal plain estuaries	Hagy et al. (2000)
ECOPATH	Energy flow in St. Francis Bay, South Africa (Kromme Estuary)	Heymans and Baird (1995)
CASM-PX	Ecosystem model applied to transect through Patuxent River estuary	Bartell et al., in review
Suspension feeding bivalve model	Ecological and hydrological model of consumption by filter feeders	Gerritsen et al. (1994)
Estuarine littoral zone model	Dynamic physical and biological model for sediments, microalgae, and SAV	Buzzelli et al. (1999)
Seagrass model	Subtropical seagrass production model, sediments, epiphytes, nutrients	Fong et al. (1997)
SSE computational model	Water and material exchange between shoals and shallow embayments, nonconservative substances, coastal plain estuaries	Kou and Park (1995)
Food web model	Model of nutrients, algae, zooplankton, fish and decomposers, nutrients, metals, risk assessment, estuaries, continental shelf ecosystems	Bartell (1994)

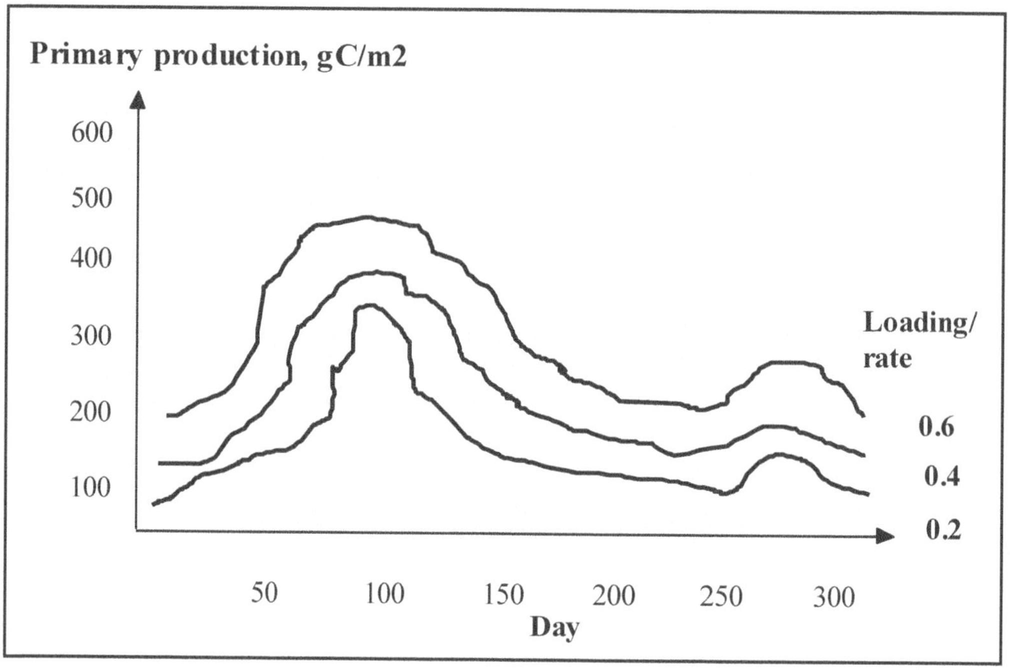

Figure 5-1. Hypothetical Modeled Daily Values of Primary Production Estimated for Fractional Increases in Natural Loading Rate of Phosphorus

Deegan et al. (1994) developed a model of nutrient flows through estuarine ecosystems (Figure 5-2). The methods of flow analysis (Finn 1976, Bartell 1978) can be used to construct a quantitative understanding of the expected response of system response variables (food web components) to unit changes in the inflow of nutrients.

Such approaches should explicitly characterize and incorporate uncertainties in estimating the ecological impacts of nutrient loadings in the context of probabilistic risk assessment (Warren-Hicks and Moore 1998, Bartell et al. 1999). Even with the dedication of resources to collect new data and information in Tiers 2 and 3, some parameter values for the selected process models that are poorly known will likely remain. Moreover, the results of statistically valid sampling may demonstrate that data describing certain estuarine model processes are inherently variable; this variability should be incorporated into the model structure and computations (Monte Carlo simulation). However, the propagation of uncertainties through the model calculations can provide statistical distributions of nutrient criteria, instead of single-value estimates. These distributions can be analyzed in relation to the model input values to identify the key parameters that control the modeled response of ecological resources to nutrient loadings (by numerical sensitivity and uncertainty analysis).

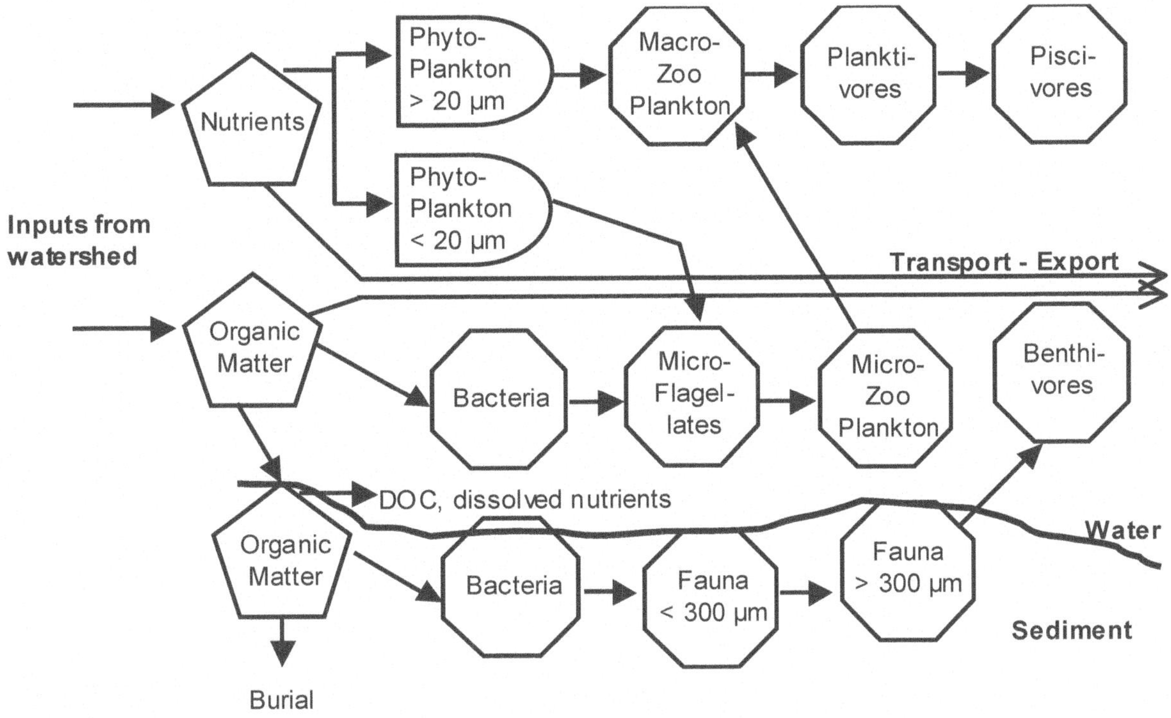

Figure 5-2. Schematic Illustration of Ecosystem Components and Carbon/Nutrient Flows Through Estuarine Ecosystem (Adapted from Deegan et al. 1994)

Where feasible, the dynamic, process-oriented models should be spatially explicit. For example, Kelly and Spofford (1977) developed a dynamic, spatial model of nutrients and associated biological production for the Delaware River and estuary. Nixon and Kremer (1977) constructed a composite simulation model to address nutrient dynamics and ecological production in the Narragansett Bay. This model was based on a finite element spatial delineation of the bay. Submodels that simulated nutrient loading, nutrient circulation, and the production of phytoplankton, zooplankton, fish, and benthos were coupled with hydrodynamic and physical-chemical submodels (light, temperature, salinity, dissolved oxygen). Hopkinson and Day (1977) developed a spatially explicit model of the Barataria Bay, Louisiana that focused on the inputs and cycling of nitrogen and the associated ecological dynamics of this estuarine system in the Gulf of Mexico. Fitz et al. (1996) describe a general ecosystem model (GEM) developed for spatially explicit application to the Patuxent River estuary to examine interrelationships between hydrodynamics, nutrient cycling, production, and decomposition in a grid-based representation of upland, wetland, and shallow-water habitats. Bartell (in review) adapted the comprehensive aquatic systems model, CASM (Bartell et al. 1999), to examine the combined effects of nutrient enrichment and toxic metals for stations that defined a transect along the mainstem of the Patuxent River and estuary. These and other estuarine modeling approaches should be reviewed and examined for potential application to Tier 3 assessments.

In a hypothetical Tier 3 assessment of SSNC for the Neuse River estuarine location near Thurman, N.C., the local stakeholders might determine that the inadequacies of the Tier 2 empirical regression models necessitated a Tier 3 evaluation. Following a comprehensive review of available estuarine ecosystem models of nutrient and production dynamics, a site-specific modification of the Narragansett Bay model (Nixon and Kremer 1977) might be selected and applied to the Neuse River estuarine site (Figure 5-3).

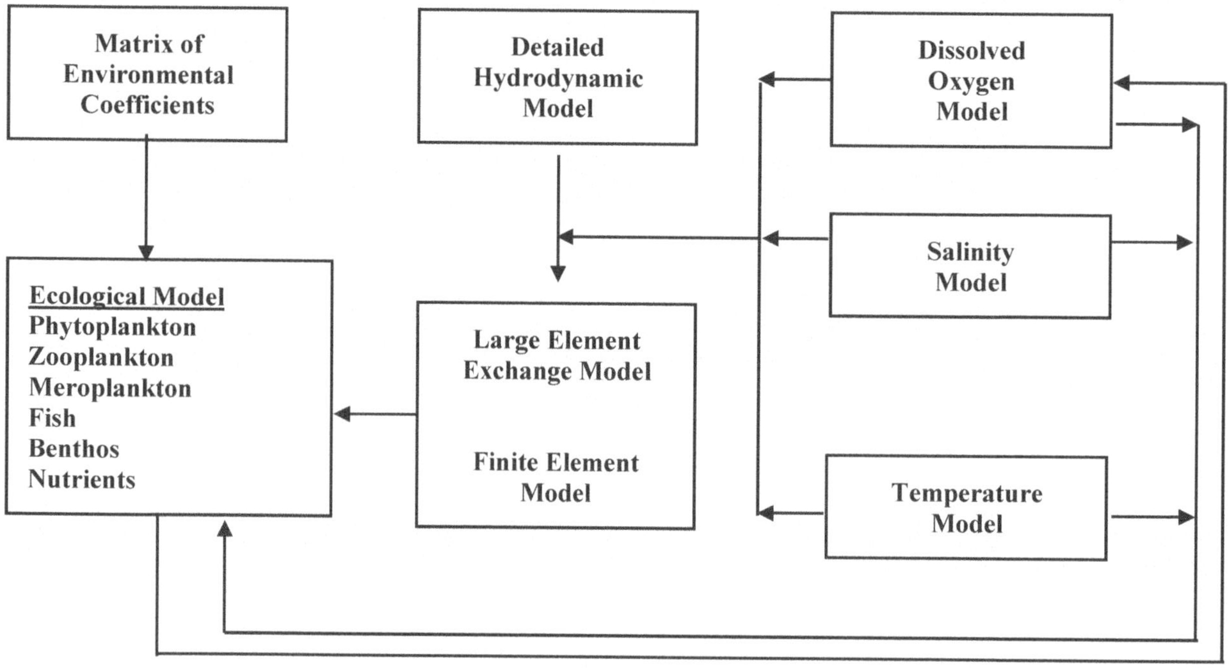

Figure 5-3. Submodels and Their Interrelationships for Narragansett Bay Estuarine Management Model (Modified from Nixon and Kremer 1977)

In theory, an augmented data base would be used to define necessary model inputs, including detailed bathymetry, temperature, and salinity data used by the hydrodynamic submodel to characterize spatial-temporal variations in current velocity and water circulation for the modeled section of the Neuse River estuary. The data base would also establish initial biomass values for the trophic guilds included in the ecological submodel. Model parameters that defined the growth dynamics of the populations of aquatic organisms would be developed from the peer-reviewed technical literature and, where possible, from the augmented Tier 3 data base. A list of model parameters, their values, the source of these values, and a quantitative or qualitative estimate of each parameter's accuracy and precision could be constructed and added to the Thurman estuary Tier 3 data base.

Continuing this hypothetical conceptual modeling exercise, the modified Narragansett Bay model would be used to examine the implications of existing and alternative future N and P loading rates to this area of the Neuse River on expected values of phytoplankton biomass (converted to Chl *a*) and the biomass of modeled benthic invertebrates (clams, shrimp, oysters) and fish (striped bass). Simulations would produce daily estimates of biomass, concentrations of dissolved N and P, and dissolved oxygen for the alternative N and P loading rates. These values would be used to establish quantitative relationships between N and P loading and Chl *a* values for the estuary. These relationships reflect the physical, chemical, biological, and ecological structure and processes formulated in the model. The relationships could be used to evaluate the Tier 1 and Tier 2 SSNC derived for TN and TP concentrations in this area of the Neuse River.

The simulations would be repeated using various input parameter values determined by assessing parameter uncertainties as listed in the Tier 3 data base. The number of model parameters, the comparative paucity of site-specific data, and the overall complexity of the model could preclude developing statistical distributions for model parameters and using Monte Carlo simulation methods. Instead, sets of "scenario-consequence" simulations should be performed, where, for each scenario of N and P loadings, model calculations would be conducted using parameter values that were systematically varied over their possible or likely range of values (interval analysis). This approach would yield information on the possible variability in model results. Most likely values and ranges of TN and TP concentrations that appeared consistent with the modeled values of phytoplankton biomass would be developed from these scenario-consequence simulations. Although not statistically rigorous, the resulting variability in model results could be used as part of the decision-making and criteria selection process.

5.6 Tier 3 Decision-making Process

The Tier 3 analyses would provide the following possible decisions concerning SSNC:

1. The selected model is successfully implemented for the system of interest. Model:data comparisons are of acceptable accuracy and precision from the perspective of the stakeholders. The SSNC can be developed using the mechanistic understanding provided by the model. Through carefully designed simulations or calculations it may be possible to use the model to directly estimate the SSNC or perhaps distributions of SSNC.
2. The model is not satisfactorily applied. Model calculations are not sufficiently accurate or precise to justify using the results or the model to derive the site-specific criteria. However, detailed sensitivity analysis of the model performance can identify key model parameters that must be better estimated to improve the model performance. The additional necessary data can be collected, the model performance can be improved, and the site-specific criteria can be developed using the model results.
3. Model performance is not improved, despite the collection of additional data, so the default, Tier 1 site-specific nutrient criteria is used. Alternatively, the possibility of selecting a different model is examined, key experiments are performed, or an intensive monitoring program is set up to provide the understanding necessary to derive the SSNC.

5.7 Selection of Tier 3 Site-specific Criteria

The first step in developing site-specific nutrient criteria using the Tier 3 results is the definition of levels of Chl *a* (or other key response variables), for example, that may be appropriate for the site and consistent with the specific water quality objectives. Based on acceptable values of Chl *a* and other selected response variables, the results of the model application can be used to estimate the nutrient loading rates and subsequent concentrations that are consistent with these values.

5.8 Implementation of Tier 3 Criteria and Nutrient Management Plans

As in Tier 2, nutrient management plans would be put in place to assure that the site-specific criteria are not exceeded once they become local standards. Such plans may be developed around the concepts and methods of TMDLs, watershed management plans, discharge limits, and other regulatory or voluntary actions.

5.9 Evaluation of Tier 3 Criteria

As in Tier 2, a monitoring program would be undertaken to ensure the attainment of the site-specific standards and that the standards adequately protect the designated uses for the segment of interest.

CHAPTER 6.0

HINKSON CREEK CASE STUDY

6.1 Hinkson Creek, Missouri–Tier 1

Hinkson Creek is a third-order stream that originates near Hallsville about 10 miles northeast of Columbia, Missouri. It flows through Columbia and into Perche Creek, about five miles southwest of Columbia. Hinkson Creek is located in U.S. EPA Ecoregion IX. U.S. EPA (2001) published 25th percentile values for TN, TP, seston Chl *a,* and periphyton Chl *a* for streams in Ecoregion IX and Level III Ecoregion 40 streams, which is the subunit of Ecoregion IX that includes Hinkson Creek (Table 6-1). For Ecoregion IX, 25th percentile TN, TP, seston Chl *a*, and periphyton Chl *a* values were 692 µg/L, 37 µg/L, 2.3 µg/L and 20 mg/m^2 respectively. For Level III Ecoregion 40, the values were 712 µg TN/L, 93 µg TP/L, and 2.8 ug Chl *a*/L. No periphyton values were published by U.S. EPA for Level III Ecoregion 40. According to U.S. EPA (2000a) guidance, these 25th percentile nutrient concentrations can be used as ecoregional nutrient criteria for Hinkson Creek.

Table 6-1. Ecoregional and Hinkson Creek 25th Percentile Values for TN, TP, and Chl *a*

Source	TN (Fg/L)	TP (Fg/L)	Seston Chl *a* (Fg/L)	Periphyton Chl *a* (mg/m^2)	TSS (mg/L)
U.S. EPA Ecoregion IX Streams	692	37	2.3	20	Not available
U.S. EPA Level III Ecoregion 40 Streams	712	93	2.8	Not available	Not available
Hinkson Creek, Missouri, 2/95 - 1/96	310	35	1.6	12	5

For illustrative purposes of this case study only, nutrient concentrations on average were assumed to be about equal to the ecoregional criteria, with occasional filamentous green algae observed. Thus, the decision was made whether a more stringent site-specific nutrient standard should be set for the creek.

6.1.1 Selecting Important Variables

For Tier 1, TN and TP were the selected nutrient variables. Chl *a* and TSS, a measure of water clarity were the response variables selected for this analysis.

6.1.2 Establishing Data Base

Daily water quality and flow data were collected from the creek by the University of Missouri personnel (Jack Jones, personal communication) from February 6, 1995, to January 31, 1996. During this period, the average TN concentration was 690 µg/L, TP was 109 µg/L, Chl *a* was 7 µg/L, benthic Chl *a* was 33.4 mg/m^2, and TSS was 71 mg/L. (Table 6-2). The number of samples taken for TN, TP, Chl *a*, and TSS was about 350 and for benthic Chl *a*, 52. Flow ranged from virtually zero to almost 300 m^3/sec, with a median of 0.067 m^3/sec. For the purposes of this case study, the assumption was made that a river survey was done evaluating relationships between peoples' perceptions of river water condition, suitability for recreation, and nutrient and Chl *a* concentrations.

Table 6-2. TN, TP, Chl *a*, Benthic Chl *a*, and TSS, Hinkson Creek, MO, 1995-1996

	TN (Fg/L)	TP (Fg/L)	Chl *a* (Fg/L)	Benthic Chl *a* (mg/m^2)	TSS (mg/L)
Mean	690	109	7	33.4	71
Median	510	55	5.6	24.3	14
Minimum	150	17	0.2	0.03	1
Maximum	2,700	1,597	57.5	163.3	1,825
Sample size	357	357	356	52	354

6.1.3 Analyzing Data

The first step in the data analysis was to plot the data to discern temporal trends. Figure 6-1 shows seasonal trends in sestonic Chl *a*, TP, and TN in Hinkson Creek from February 1995 through January 1996. The highest Chl *a*, TP, and TN concentrations were present during the growing season from about mid-April through mid-September. Both the annual average and spring and summer TN:TP ratios were slightly less than nine, indicating that Hinkson Creek probably is P limited, but also may be N limited at times. About 31% of samples had ratios between 10 and 20, indicating that phosphorous is limiting phytoplankton growth at times.

The second step in the data analysis was to evaluate nutrient response variable relationships by plotting Chl *a* versus TN and TP. Both TN and TP appear to be positively related with Chl *a*, although the relationship is quite variable (Figure 6-2). Both TN and TP also were positively correlated with flow (Figure 6-3). The highest TN and TP concentrations were associated with the highest stream flows, suggesting that watershed runoff during precipitation events was the source of the high TN and TP inputs. Comparison of TN and TP with Chl *a* during the growing season showed no significant relationships, possibly suggesting a delayed response between the input of nutrients and phytoplankton growth. These comparisons indicate that reductions in both TN and TP should lead to reductions in Chl *a*.

The third step in the data analysis was to develop cumulative frequency distributions for each nutrient and response variable. Figures 6-4 through 6-8 are cumulative frequency distributions for measurements of TN, TP, seston Chl *a*, periphyton Chl *a*, and TSS in Hinkson Creek, Missouri. The data are based on samples taken from February 6, 1995 through January 31, 1996.

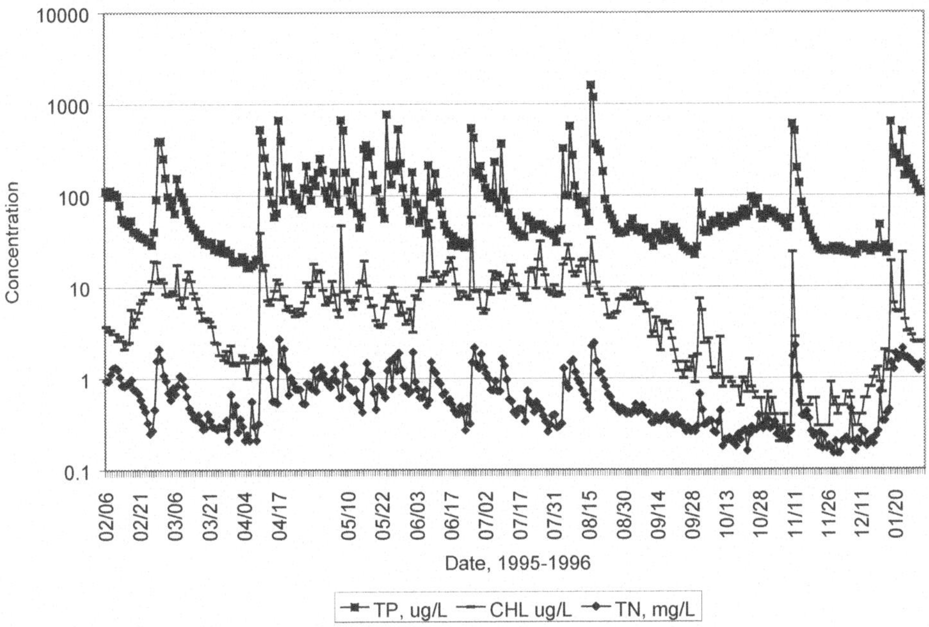

Figure 6-1. Seasonal Trends in Sestonic Chl *a*, TP, and TN in Hinkson Creek near Columbia, MO, February 1995-January 1996

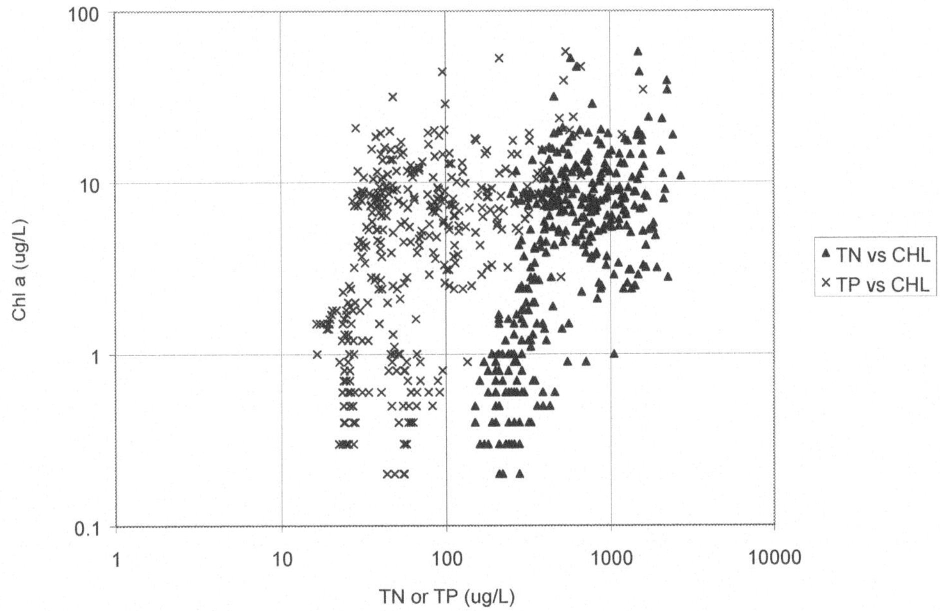

Figure 6-2. Relationship Between Sestonic Chl *a* and TN and TP in Hinkson Creek near Columbia, MO, February 1995-January 1996

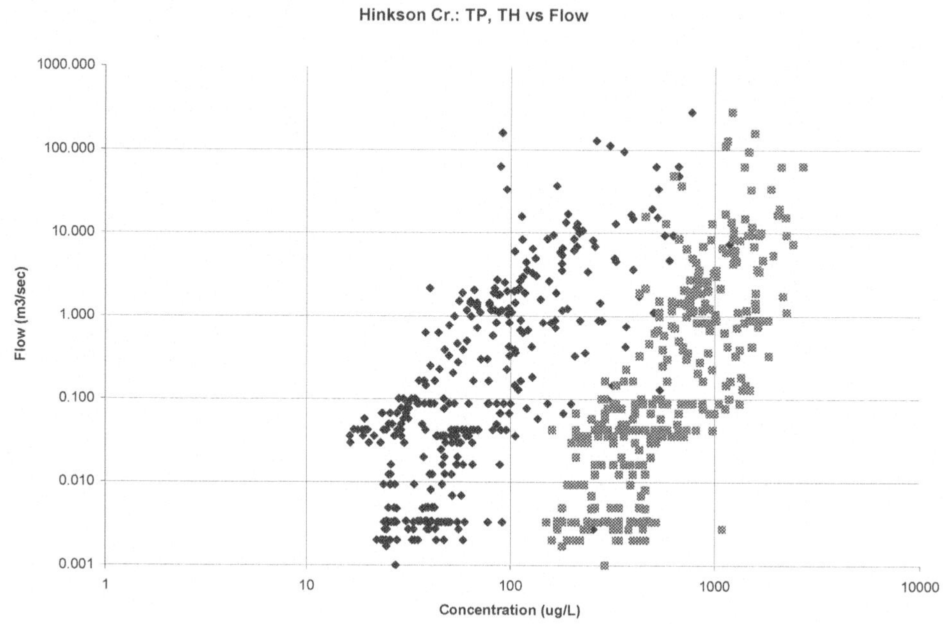

Figure 6-3. Relationship Between Flow and TN and TP in Hinkson Creek near Columbia, MO, February 1995-January 1996

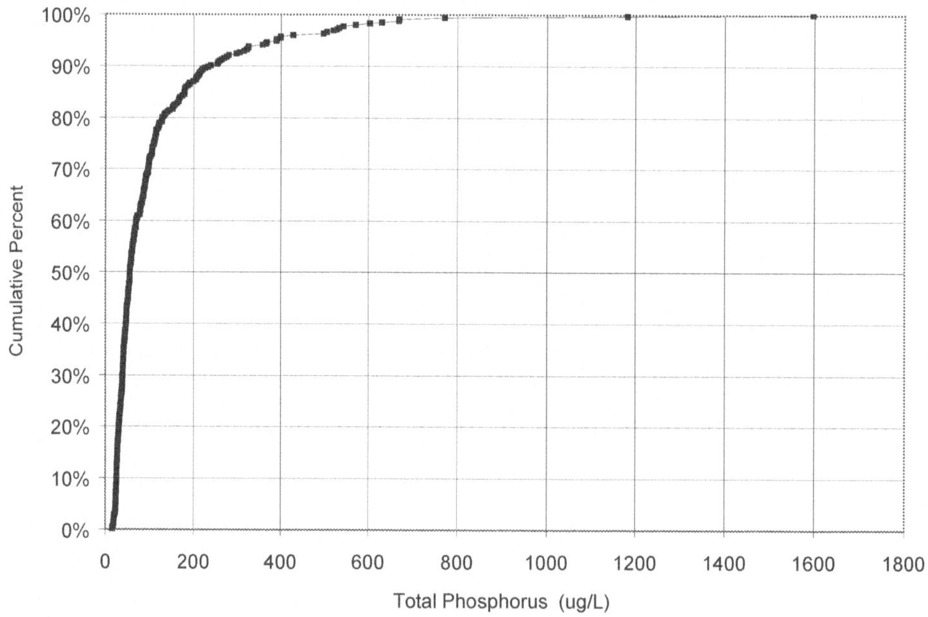

Figure 6-4. Cumulative Distribution of TP in Hinkson Creek near Columbia, MO, February 1995-January 1996

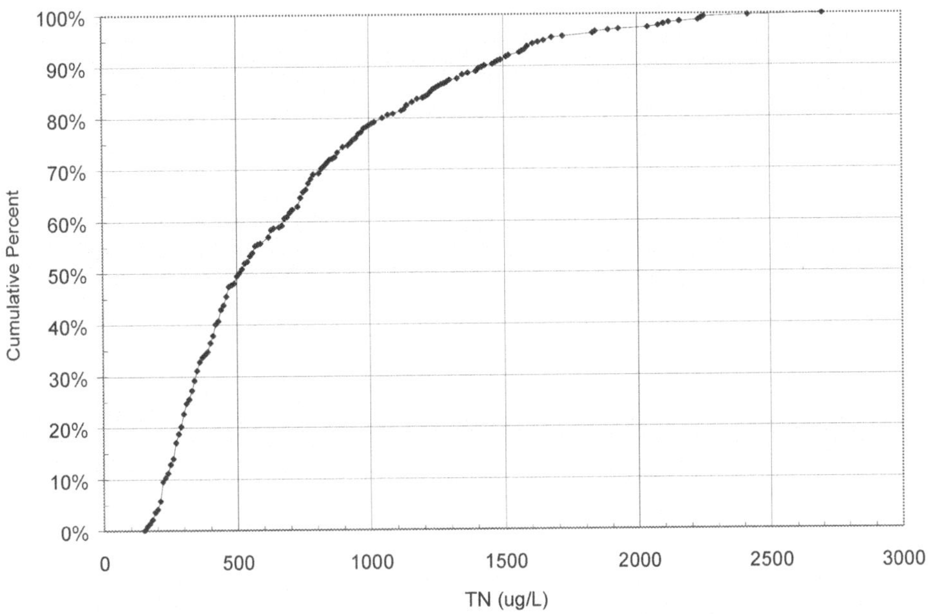

Figure 6-5. Cumulative Distribution of TN in Hinkson Creek near Columbia, MO, February 1995-January 1996

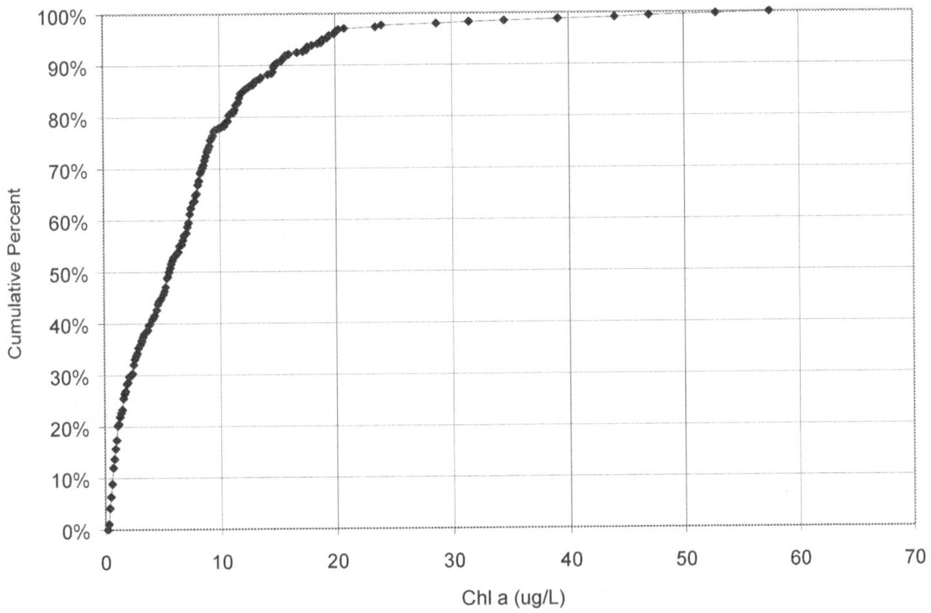

Figure 6-6. Cumulative Distribution of Sestonic Chl *a* in Hinkson Creek near Columbia, MO, February 1995-January 1996

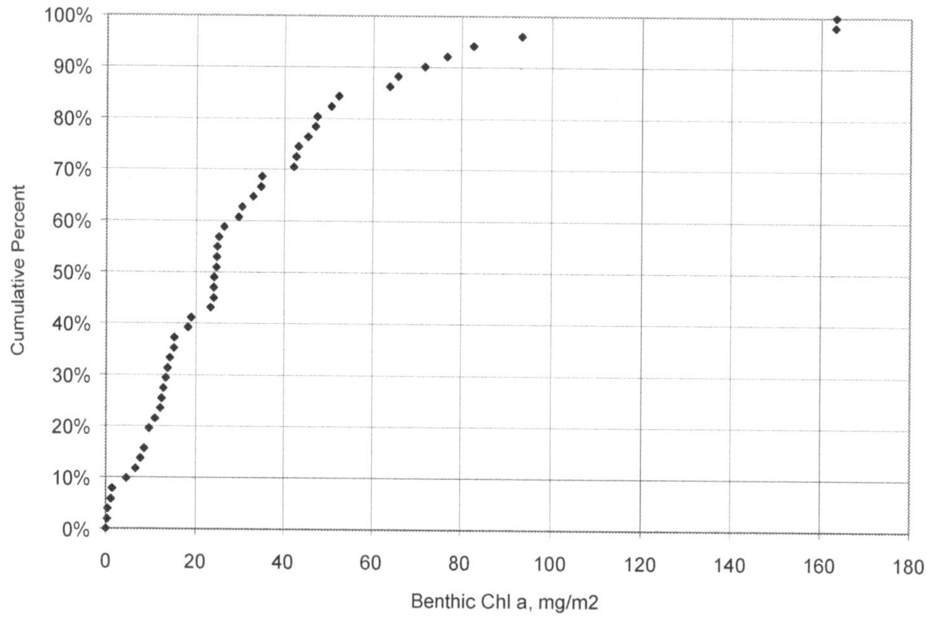

Figure 6-7. Cumulative Distribution of Periphyton Chl *a* in Hinkson Creek near Columbia, MO, February 1995-January 1996

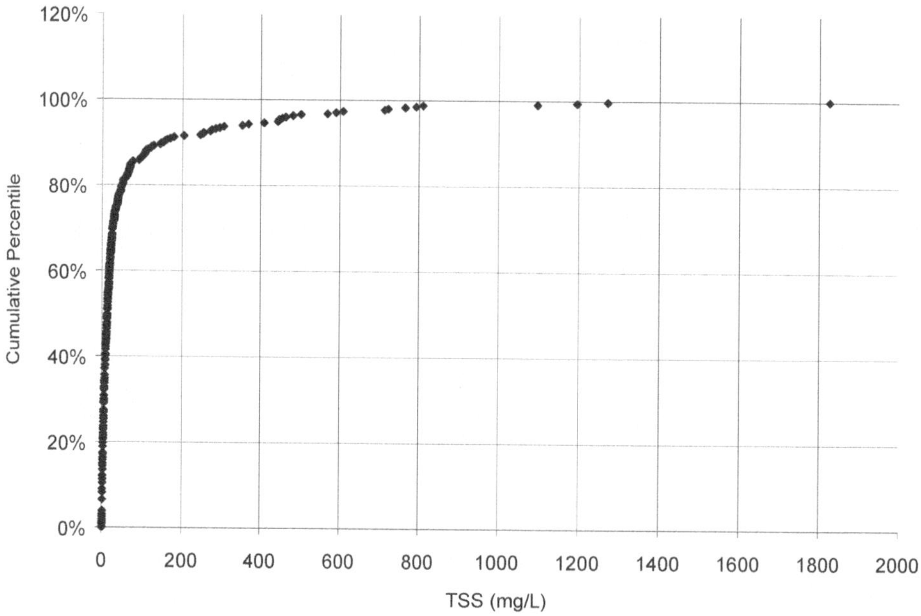

Figure 6-8. Cumulative Distribution of TSS in Hinkson Creek near Columbia, MO, February 1995-January 1996

6.1.4 Selecting Criteria

The lower 25th percentile values were 310 µg/L, 35 µg/L, 1.6 µg/L, 12 mg/m^2, and 5 mg/L for TN, TP, sestonic Chl *a*, benthic Chl *a*, and TSS (Table 6-1), respectively. The upper 75th percentile values were 925 µg/L, 109 µg/L, 9.2 µg/L, 43 mg/m^2, and 89 mg/L, respectively. These or other values could be selected as Tier 1 SSNC, if they were found to be associated with acceptable water quality conditions, such as Chl *a*, algal blooms, drinking water taste and odor, and attainment of designated uses, and did not adversely affect downstream uses. These values indicate the relative disparity of results that could occur using the ecoregional approach, leading to the conclusion that the ecoregional approach may not be appropriate for site-specific standards. The results of a stream survey could be used to select the most appropriate site-specific standard. The site-specific 25th percentile values for Hinkson Creek were less than both the Ecoregion IX and Level III Ecoregion 40 values (Table 6-1), while the site-specific 75th percentile values were greater than the ecoregional values. To be the most protective of water quality, the lower 25th percentile values for Hinkson Creek could be selected as the SSNC.

6.1.5 Implementation and Management

If the proposed lower 25th percentile values were acceptable to regulatory and regulated entities, these proposed concentrations would be submitted to the state water quality management agency for implementation. Nutrient management plans would be implemented (TMDL determinations, BMPs, discharge limits).

6.1.6 Long-term Monitoring and Evaluation

If the proposed Tier 1 SSNC were accepted, water quality would continue to be evaluated with an ongoing monitoring program to ensure that criteria and designated uses were being met. In this case study, the assumption was made that Tier 1 SSNC were not implemented and that further study and review would be required, resulting in a Tier 2 analysis.

6.2 Hinkson Creek—Tier 2

6.2.1 Review and Revision of Variables

In addition to the Tier 1 variables, other variables were evaluated in Tier 2. These variables included those that might have confounded relationships between nutrients and Chl *a* such as streamflow and TSS.

6.2.2 Augmenting Data Base

In Tier 2, data sources are reviewed and if needed additional data collected. One year of data on Hinkson Creek was determined to be sufficient, and thus no additional data were collected.

6.2.3 Conducting Empirical Data Analyses

The relationships between TN, TP, and sestonic and benthic Chl *a* were further investigated. A significant relationship between daily TN and sestonic Chl *a* was observed (Figure 6-9) throughout the year, but no such relationships were apparent for data from the growing season. The R^2 for the relationship for the entire year was 0.43, indicating that TN accounted for about 43% of the variability in Chl *a*. By fitting a cubic, polynomial model to the data, the R^2 increased to about 0.55 (Figure 6-10). The R^2 for the relationship between TP and Chl *a* was only 24% (Figure 6-11). The R^2 for the multiple regression between TP, TN and Chl *a* also was 43%, but in this equation the coefficient for TP was negative, indicating that the variability in TN wass dominating the relationship. Figure 6-3 also indicates that concentrations of TN and TP in Hinkson Creek were correlated. Figures 6-9 and 6-10 suggests that a threshold for the response of Chl *a* to TN occurred in Hinkson Creek. Sestonic Chl *a* increased as TN increased in the creek up to about 300 to 400 µg TN/L, after which Chl *a* concentration did not increase. Above this concentration, TN appeared to be present at levels that exceeded needs by algae. A similar threshold was apparent for TP and Chl *a* (Figure 6-11).

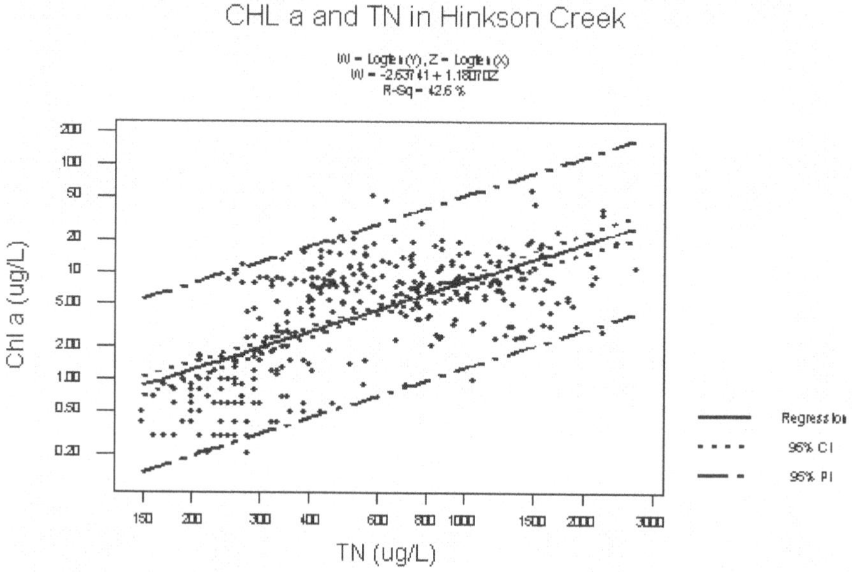

Figure 6-9. Linear Regression Model of Sestonic Chl *a* and TN in Hinkson Creek near Columbia, MO, February 1995-January 1996

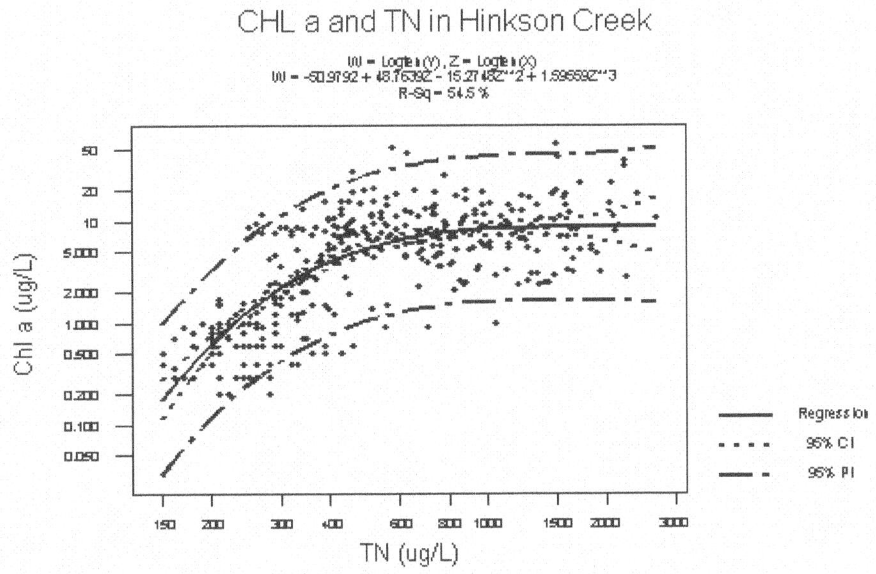

Figure 6-10. Cubic Model of Sestonic Chl *a* and TN in Hinkson Creek near Columbia, MO, February 1995-January 1996

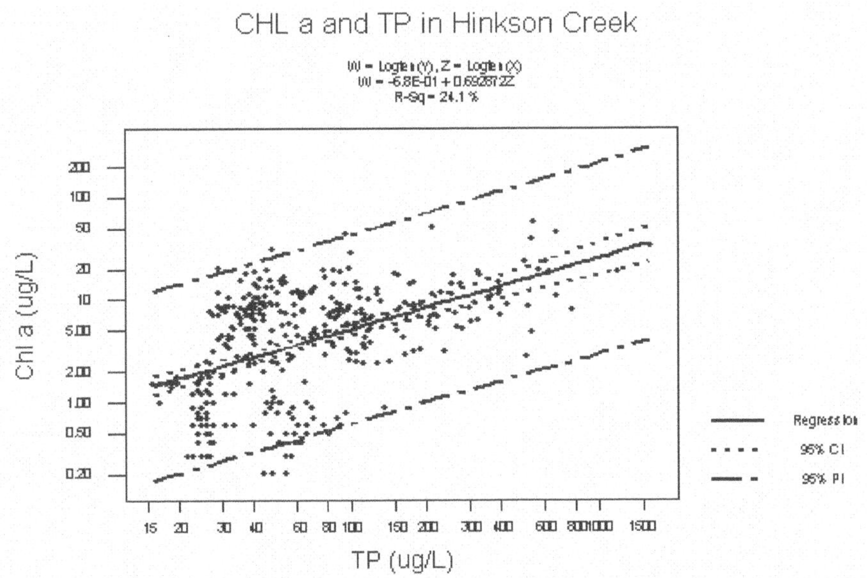

Figure 6-11. Linear Regression Model of Chl *a* and TP in Hinkson Creek near Columbia, MO, February 1995-January 1996

No significant relationships were found between TN, TP, or SRP, and periphyton Chl *a*, which precludes using these data for development of Tier 2 SSNC. Nonvolatile suspended solids were related to Chl *a*, though only weakly ($R^2 = 0.28$), suggesting that turbidity might be confounding the relationship between Chl *a* and nutrients. Flow and TSS also were weakly related ($R^2 = 0.30$), suggesting that runoff might be a source of TSS.

6.2.4 Selecting Site-specific Criteria

The significant relationships between TP, TN, and sestonic Chl *a* could be used to set SSNC. For example, Figure 6-10 suggests that TN concentrations in Hinkson Creek would have to be reduced to less than 300 to 400 μg/L before reductions in sestonic Chl *a* would occur, although the uncertainty in the relationship is large as shown by the 95% confidence and prediction intervals. If, for example, a SSNC Chl *a* concentration of about 2 μg/L was set as the water quality goal for Hinkson Creek (using the model in Figure 6-9), TN in the creek would have to be reduced to a maximum of about 350 μg/L to achieve that goal. The uncertainty in the relationship also suggested that TN might have to be reduced even more, possibly to as low as about 200 μg/L to achieve a Chl *a* concentration of 2 μg/L.

The SSNC for TP also could be established using the correlation between Chl *a* and TP, or the TN:TP ratio. Based on the Chl *a* and TP relationship shown in Figure 6-11, if the SSNC for Chl *a* were 2 μg/L, TP would have to be reduced to a maximum of about 30 μg/L. Interestingly, the SSNC for TN and TP developed from these regression modelweare relatively close to those based on 25th percentiles for Hinkson Creek data. One advantage of the empirical model-based SSNC compared to those developed from the 25th percentiles is that the model-based SSNC provides information on the potential uncertainty in the response of Chl *a* to reductions in concentrations of TN and TP. For example, at a TN concentration of 300 μg/L, Chl *a* ranged from 0.4 μg/L to almost 10 μg/L. A key environmental factor that might be contributing to this uncertainty was streamflow, which is correlated with both TN and TP concentrations (Figure 6-3).

6.2.5 Implementation and Management

If the proposed concentrations were acceptable to regulatory and regulated entities, they would be submitted to the state water quality management agency for implementation. Nutrient management plans would also be developed and implemented (TMDL determinations, BMPs, discharge limits). As discussed previously, TN and TP concentrations were positively related to flow in Hinkson Creek, and 57% of the variability in TN was explained by flow, suggesting that runoff was a major source of both nutrients. If true, efforts to reduce TN and TP in Hinkson Creek to achieve reductions in Chl *a* would have to focus on reducing runoff.

6.2.6 Long-term Monitoring and Evaluation

Water quality in Hinkson Creek would continue to be evaluated with an ongoing monitoring program to ensure that criteria and designated uses were being met.

APPENDIX A

A Framework for Stakeholder Involvement in the SSNC Process

A.1 Identifying Stakeholders

Stakeholders include all agencies, organizations, and individuals involved in or affected by nutrient water quality management decisions. Stakeholders may include:

- local government agencies,
- local offices of federal agencies,
- local water quality research staff,
- NPDES dischargers and water supply users,
- environmental groups,
- business groups,
- the general public,
- appointed natural resource trustees,
- state water quality managers, and
- regional offices of the U.S. EPA.

Stakeholders play an important role in the overall process for developing SSNC. An active stakeholder process can ultimately lead to better, more cost-effective water quality management decisions. Cost-effective water quality management involves efficient and timely analysis, an understanding of water quality objectives, the ability to prioritize the objectives, and the allocation of resources to water quality management activities that will most effectively satisfy stakeholders' concerns. For site-specific nutrient criteria to be an integral part of cost-effective water quality management, stakeholders should be involved in several ways.

- Stakeholders should be included in early and frequent communications with the water quality assessment team and the water quality managers.
- Stakeholders' values and concerns should be described and understood at each step in the SSNC process.
- Stakeholders should be provided with a working knowledge of water quality assessments and the process for setting water quality criteria and standards.
- Stakeholders should participate in selecting those potential nutrient issues or problems that become the subject of time-consuming, detailed, and expensive analysis.

- Stakeholders should be made aware of the context for SSNC assessments in the water quality management process—who makes the final decision and how the water quality assessment team and the stakeholders influence decisions.

A.2 Stakeholder Involvement in the Site-specific Nutrient Criteria Process

A.2.1 Opportunities for Participation

There are at least four opportunities for stakeholders to actively participate in the derivation of SSNC:

1. specifying the degree of stakeholders' involvement,
2. affecting the perception of water quality problems caused by nutrients,
3. formulating nutrient water quality objectives, and
4. providing the context for water quality assessment.

Water quality objectives can be viewed objectively, subjectively with personal bias, or at some intermediate level. People's environmental values may be fixed and unwavering, or flexible and amenable to refinement given new information, education, or context. The context of the assessment or the set of questions the process seeks to answer may deal exclusively with narrow regulatory requirements or it may broaden to issues not addressed directly by the regulations. While most stakeholder processes address this topic to varying degrees, the SSNC process is designed to make the most effective use of the stakeholders' involvement in the derivation of SSNC.

A.2.1.1 Stakeholder Involvement

How much stakeholder involvement is enough? Involving stakeholders has traditionally been limited to advertising the public's opportunity to review and comment on draft reports that summarize scientific analyses and recommendations. This approach stems from the typical requirements of current regulations. More recently, people recognize that advisory groups, comprising diverse members of the community and already established for some watersheds or waterbodies to provide guidance on managing environmental issues, can make valuable contributions to the SSNC process. In this broader role, the advisory group provides input on nutrient concerns at the beginning of the process and then assists in reviewing and evaluating alternative SSNC after the assessment is completed.

The SSNC stakeholder process is designed to involve stakeholders actively at every stage of the process, going well beyond simply eliciting of comments on the final SSNC report or product. For example, the advisory group can obtain formal feedback from the constituencies it represents by hosting community-wide forums or conducting surveys to elicit opinions on nutrient-related water quality objectives. Determining how much stakeholders' involvement is adequate depends on external factors, such as the nature of the decision to be made (is the issue straightforward, controversial, or complicated?) and the community's degree of interest in the issue.

A.2.1.2 Perception of Nutrient Water Quality Problems

Are water quality problems caused by nutrients viewed as objective (products of the scientific process) or are they viewed as subjective (the products of individual perceptions and values)? Depending on how stakeholders participate, nutrient water quality problems are viewed in various ways. One stakeholder participation model operates on the principle that a scientific (completely objective) exercise is required to evaluate nutrient water quality problems. Scientists convey the magnitude and severity of problems quantitatively to stakeholders and water quality managers, so they can evaluate options for addressing them. Communication with the stakeholders is essentially one way and begins when the scientific process is completed.

The underlying premise of a contrasting model is that nutrient water quality problems are subjective and can be classified as perception. The conflict between Aactual@ and Aperceived@ nutrient water quality problems is really the difference between two sets of water quality perception: those of scientists working in their field of expertise and those of everyone else (Fischoff 1990). In this model, stakeholders work as partners with scientists at every stage of the process—defining the problem to be studied, prioritizing the water quality objectives, and evaluating and ranking the solutions. Communication is two way and iterative, ensuring that the SSNC process clearly and fully addresses the stakeholders' ecological values.

In contrast to the first model, this second model views water quality management as an integral part of the SSNC process rather than a separate step. In this approach, water quality assessors must particularly attend to Aframing effects@ and avoid influencing the responses of stakeholders with the way information is presented and values elicited. For example, the order of topics in a survey can affect the range and mixture of items being evaluated; the amount and nature of information provided about each item, the method of elicitation, and many other contextual factors can also affect the results of the survey (Gregory and Lichtenstein 1993).

A.2.1.3 Value Formulation

Do environmental values for water quality exist or must they be constructed? Are people's values fixed or are they flexible, given new information or a different context? The "archeological" approach to ascertaining stakeholders' environmental values and concerns requires that water quality values exist and that they can be revealed by a survey or other means.

The architectural approach focuses on the multi-dimensional nature of environmental values and on the constructive nature of human preferences (Gregory and Lichtenstein 1993). Under this approach, working with stakeholders is an active process that involves "value construction" rather than a neutral process that relies merely on "value discovery."

The SSNC stakeholder process approach to identifying stakeholders' values and concerns differs considerably from the usual approach. The SSNC process assumes that while people's basic environmental beliefs are strongly held and generally unwavering, individuals may alter their values and concerns when presented with new information (or known information set in a different context). The SSNC process provides a way for people to draw upon their basic beliefs when asked to quantify and rank their environmental preferences. In this regard, the approach to stakeholder involvement is distinct from other such paradigms. Elements of the architectural approach merge with archaeological elements to achieve a clear understanding of stakeholders' values and how they are ranked. Promoting an open-minded exchange both among stakeholders and between stakeholders and the water quality assessment team, the SSNC approach encourages stakeholders individually and collectively to quantify and rank relative environmental values for nutrient water quality, while attempting not to change stakeholders' fundamental tenets regarding the environment.

A.2.1.4 Context of the Assessment

How do we determine the spectrum of water quality problems caused by nutrients being evaluated? In one model, federal regulations, policy, and guidance define the scope of water-quality assessment. In the SSNC approach, the definition and comparison of nutrient water-quality objectives are restricted, addressing only those objectives that are the direct subject of the applicable regulation or guide. A second model addresses water quality within a broader context, comparing related, regulatory, and nonregulatory parameters, such as those that adversely affect primary production, or the potential adverse effects of reductions in nutrients on fish production. The SSNC stakeholder process addresses the broader context of water quality assessment, considers the effects of a variety of physical, chemical, and biological factors on nutrient-algal relationships, and is limited by those water quality regulations. Cost-effective water quality management requires participants to compare water quality objectives, including some beyond the often narrow scope of regulations.

A.2.2 The SSNC Stakeholder Process Philosophy

Many stakeholder processes focus on one aspect of stakeholder involvement (for example, perception of nutrient water quality objectives, value formulation, and context). The SSNC process attempts to blend science, meaningful stakeholder involvement, and cost-effective water quality management. Table A-1 outlines the core philosophical principles that shape the SSNC stakeholder process, and Section A.2.3 describes the approach to water quality assessment and how stakeholders are involved throughout the process.

Table A-1. Core Principles Shaping the SSNC Stakeholder Process

1.	Stakeholders, water quality managers, and water quality assessors maintain effective communication and mutual respect throughout the process.
2.	Establishing realistic expectations about the role of stakeholders and describing how their input will be used is essential.
3.	Stakeholders' ecological values and concerns are integral aspects in defining the water quality assessment problem.
4.	The water quality assessment process is both objective and subjective.
5.	Working with stakeholders is an active process that helps translate general environmental values into specific concerns and assists in ranking concerns through a comparative valuation process.
6.	Cost-effective water quality management requires that people be able to compare water quality objectives, including those beyond the regulations driving the water quality assessment process.
7.	Sound science, including documenting uncertainty, key assumptions, and the role of best professional judgment, is the foundation for water quality management.
8.	The statistical, quantitative estimate of water quality is only one factor upon which decisions are based; nonquantitative factors also frame water quality management decisions.

A.2.3 Stakeholder Involvement with the SSNC Process

Stakeholders become initially involved in the SSNC process by participating in an educational forum. This forum (described in Section A.3.3) accurately defines the meaning of scientific concepts, terminology, and procedures to stakeholders who might have no previous experience with the process. Once informed, the stakeholders continue by 1) identifying values and concerns that should be addressed in the water quality assessment, 2) reviewing nutrient and biological response variables to be evaluated in the SSNC process, and 3) participating in a site-specific survey to evaluate relationships between stakeholder values, concerns, and nutrient and biological response variables.

Stakeholders remain involved throughout each tier of the SSNC process to clearly identify and thoroughly address the community's environmental values. Stakeholders continue evaluating SSNC developed in the previous tier and provide input on the need for additional data. In Tiers 2 and 3, stakeholders may participate in SSNC ranking. Stakeholders also may play an integral role in reviewing data gaps and potential yet nonquantifiable nutrient water quality problems, the need for future monitoring, data collection, and research.

The challenge lies in balancing meaningful stakeholder participation with progress in water quality management. Rather than allowing the stakeholders to access the water quality assessment team and process at all times, the SSNC process strikes the participation/management balance by providing well-defined, strategic time frames for stakeholder input.

A.3 Involving Stakeholders in the SSNC ProcessCGetting Started

A.3.1 Preliminary Steps

The SSNC process begins when a waterbody is identified as a candidate for SSNC. Typically, this occurs because water quality monitoring data collected by a regulatory or nonregulatory entity have indicated that existing ecoregional nutrient water quality criteria are not being attained in the waterbody; nutrient water quality problems, such as nuisance algal growths or fish kills, have been observed in the waterbody; or large expenditures will be required to meet ecoregional criteria in a waterbody that already is attaining designated uses. As a result of these types of occurrences, a stakeholder, such as a municipality with a wastewater treatment plant, a water quality management agency, a lake homeowners' association, or an individual, may propose that either less stringent or more stringent nutrient criteria are appropriate for the waterbody. The formal stakeholder process should begin at this point. To facilitate the SSNC process, it would be beneficial if the U.S. EPA regional office, the state water quality management agency or a nongovernmental organization (NGO) had a formal mechanism in place for initiating the SSNC stakeholder process. For example, the mechanism could include a procedure by which the stakeholder could petition the state water quality management agency or NGO to begin the SSNC process. In the petition, the stakeholder would briefly describe the waterbody, the nutrient water quality concerns or issues, and the likely benefits of SSNC. If the petition were granted, the stakeholder process would begin by establishing a formal stakeholder group.

After the stakeholder group has been established, its first task would be to define the aquatic resources of concern and the nutrient water quality objectives for the waterbody. These objectives may include: 1) formal, regulatory designated uses identified in state water quality standards, such as the support of warm water fisheries, drinking water, and full body contact; 2) numeric water quality standards, such as those for nutrients, Chl *a*, DO, and turbidity; 3) narrative water quality standards, such as "the waters must be free from nuisance growths of algae"or "the waters must not be chemically toxic"; or 4) nonregulatory objectives, such as the maintenance of a productive fishery for smallmouth bass or the absence of noxious weed growths.

The initial steps of the stakeholder process can greatly influence its outcome. First, an organizational structure must be established, including a neutral facilitator, water quality assessment team, administrative staff person, and a technical advisory group (Figure A-3). Second, before advisory group members begin "advising," the entire group should be provided with a common base of information to facilitate future communication.

A.3.1.1 Managing the Process

Regulatory officials should select a neutral facilitator to help design and manage the stakeholder process and an administrative staff person to organize meetings, handle nontechnical issues, and ensure that procedures are consistent with regulatory policies. Regulatory officials should also consider selecting a neutral water quality assessment team, which would identify appropriate regulatory personnel to serve as a technical support group. This group would organize, analyze, and interpret existing data; participants in investigations previously conducted in the study area would be logical candidates.

The facilitator should organize and manage the stakeholder advisory group, which may choose to operate informally or according to a more formal structure. Stakeholders may wish to adopt a two-phase approach for community participation: an advisory group of "voting" members and a group of other interested parties in the community who wish to contribute to the discussions. Regardless, defining the ground rules for the stakeholder process up front is vital—what the format will be for obtaining input and reaching agreement should be clarified, and the level of consensus required for adopting recommendations should be specified.

A.3.2 Establishing the Stakeholder Advisory Group

A.3.2.1 Recruiting Local Stakeholder Representatives

While state water quality managers and U.S. EPA officials play a significant role in the regional nutrient water quality assessment process, the SSNC process emphasizes local stakeholders and ensures that local environmental values and concerns are the major focus. Therefore, local stakeholders should constitute the official advisory group, with state and U.S. EPA representatives serving as *ex officio* members.

Potential individual members should include people:

- sufficiently interested to commit the time required for project participation;
- trusted and respected by the community; and
- open-minded and willing to work with others, even those having disparate views.

After identifying a target stakeholder pool, a concise, nontechnical project description should be developed to facilitate initial contacts with stakeholder groups and to recruit 5-10 members representing diverse local interests. Ensuring a responsive, well-rounded, and capable advisory group rests on a thorough list of the various interests requiring representation and the criteria important for selecting community representatives. This initial outreach should describe the SSNC process, identify the need for existing data and reports, and emphasize the opportunity for local stakeholder involvement. Numerous environmental studies and initiatives may have been conducted on the waterbody; therefore, conveying explicitly how this effort both builds on and differs from other efforts is critical. Why stakeholder groups would be interested in being involved and what specifically is needed from stakeholder representatives should be stressed.

Awareness of past and ongoing public participation efforts is integral to successfully implementing the stakeholder process. The SSNC process can be facilitated by capitalizing on the expertise of existing groups and drawing on previous work. Local working groups, key community leaders, members of environmental organizations, and business persons can be critical resources.

A.3.2.2 Role of the Advisory Group

Advisory group members play a challenging role encompassing many diverse responsibilities, from identifying and ranking ecological concerns of stakeholders to reviewing preliminary assessment results to recommending future data collection efforts. Specific responsibilities include the following:

- Identifying nutrient water quality objectives and concerns to be addressed in the SSNC process. For example, facilitators should identify and compile nutrient water quality objectives and attributes highlighted in recent community visioning or environmental studies. The advisory group can then help answer questions such as: Is this list complete? How do the nutrient water quality objectives and attributes rank in priority? Which should be measured and when?
- Reviewing the results of and providing feedback on the SSNC process.
- Reviewing nutrient and biological response variables and nutrient water quality management options.
- Ranking the potential nutrient water quality objectives (after the group defines them and determines their probability of occurring) that would benefit most from SSNC.
- Reviewing data gaps and prioritizing the need for future monitoring, data collection, and research.
- Providing a link to the community at large for obtaining feedback on the process.

As they rank nutrient water quality concerns, objectives, and remedies, the advisory group members can 1) formally obtain feedback from and speak for a specific interest group, 2) present their personal views, 3) inform their constituencies of project progress, or 4) help organize open community forums. The chosen approach should reflect what is likely to work in the community within the time frame of the SSNC process.

A.3.2.3 Building a Common Base of Understanding

Forging a common base of understanding among the stakeholders depends on reaching general agreement on six key elements:

1) information and education needs regarding the SSNC process;
2) overall direction and vision;
3) geographic area or community of interest;
4) methods for communicating with, and obtaining feedback from, the community at large;
5) scope of work and schedule; and
6) procedure for documenting progress and recommendations.

***Obtaining Information and Education*:** Understanding and agreeing on the definitions of terms is vital if the stakeholder process is to operate efficiently. Initiating the stakeholder process with a one- to two-hour educational session on the SSNC process can lay the foundation for a common understanding. Section A.3.3 describes the content of the education module. The education module should define the most important terms and methods that will be used in the SSNC process. Water quality assessors and group facilitators should strive to avoid jargon when working with the advisory group, but stakeholders should have access to a glossary when reading nutrient water quality literature or listening to presentations containing nutrient water quality assessment terms.

***Achieving Common Direction and Vision*:** A second, integral aspect of developing a common base of understanding promotes a common vision created by the advisory group. Toward this end, the first advisory group meeting should be used to convey the overall direction and purpose of the SSNC process, as well as the roles and responsibilities of stakeholders. Establishing the direction and roles up front is essential for three key reasons. First, achieving a common understanding of, and concurrence with, the process is paramount to moving through it smoothly. Second, creating realistic stakeholder expectations is critical and can be accomplished by clarifying how their input will be used, who is responsible for final decisions, which issues have legal or regulatory importance and constraints, and what time and budget constraints exist. Finally, the group should agree on the potential benefits of the SSNC process. During times of contention, reminding group members of their role and the numerous benefits to be gained may be one of the most effective ways to regain focus.

***Selecting the Waterbody*:** The SSNC process is not limited to the federal regulatory environment; therefore, the group must concur on the waterbody to be evaluated.

***Communicating Progress*:** Periodic communications from the group to the community at large are crucial, as is obtaining feedback. Advisory group staff should facilitate this effort by providing meeting minutes and summaries of progress.

***Determining the Scope of the Assessment*:** The advisory group should agree on the scope of work and the timetable for its completion. Formalizing the group's work plan can help achieve a common vision and maintain a sense of accomplishment and progress. A definitive work plan can also focus the direction of the group as it works together over the year of the project.

***Reporting to Water quality Managers*:** Finally, the group requires an understanding of the methods for documenting its work and recommendations.

A.3.3 Effective SSNC Process Education

A key feature of the SSNC process solicits active stakeholder involvement and provides a solid foundation for participation. The terminology and concepts used in the field of nutrient water quality assessment can be overwhelming and confusing to those not schooled in the environmental sciences. SSNC's stakeholder process begins with an educational module that introduces the key terms, concepts, and procedures.

The SSNC education module could consist of 1) a slide or overhead presentation with each graphic or bulleted text slide annotated with explanatory text; 2) an illustrative Acase study@ that sets the stage for a hypothetical nutrient water quality assessment, describing the characteristics of the waterbody; and 3) supporting reference materials, such as lists of terms used by nutrient water quality assessors. Concise, straightforward examples emphasized throughout the course can maximize the amount of information conveyed without overwhelming the stakeholders with jargon and complicated concepts.

Such a course could begin with an explanation of what is meant by technical terms such as "eutrophication" and "nutrient water quality objectives" and why SSNC are developed. Next, the types of information that result from the SSNC process, or what is learned from it, could be described. The meaning of uncertainty in nutrient water quality assessments could be explained.

Participants in the course could then learn to interpret the results of the nutrient water-quality assessments and how the underlying assumptions affect those results. Finally, the SSNC process and its three distinct tiers would be defined. The course might conclude with an illustration using the example case study.

The SSNC education module should provide stakeholders with a common understanding and a solid working foundation that enhances their participation in and contribution to the nutrient water quality assessment process.

A.4 Determining Stakeholder Values and Concerns for Inclusion in the SSNC Process

This step determines 1) which stakeholder environmental values should be measured when assessing the nutrient problems and 2) how the stakeholders value or weigh different environmental attributes. While nutrient water quality assessors may study additional attributes, they should at least address the major concerns of the community. Recent and ongoing environmental studies can contribute initial information for identifying the community's ecological values and concerns; the next source would be the results of a group survey.

A.4.1 Building on Previous Studies

As a first step, the SSNC process researches recent or current environmental studies of the waterbody that have elicited public participation or comment. This exercise explores the community's history of environmental concerns and ecological values and attempts to build on previous work. Findings should be summarized in clear, concise reports for the stakeholder advisory group.

A.4.2 Designing a Water Quality Perception Survey

Once the community's nutrient water quality concerns and values have been compiled from recent studies, the list can be refined and corroborated by conducting a water quality perception survey with the stakeholder advisory group. This step involves preparing a chart listing general nutrient water quality attributes. Under each general heading, the specific related attributes that people indicate require protection, restoration, or creation are listed. This more specific set of attributes must be measurable (parameters that nutrient water quality assessors can readily track as indicators when comparing nutrient water quality management alternatives).

At the first meeting, the advisory group should determine which environmental attributes will be measured and their relative importance to the group by:

1) refining a draft list of nutrient water quality values and attributes;
2) allocating 100 points to each member and instructing each to distribute the points across all the values and attributes;
3) summing the points in the general categories;
4) having members check their point spread across major categories to determine whether the spread reflects relative importance and, if not, asking them to readjust points;
5) compiling the results and discussing them as a group to determine whether major differences among values occur;
6) asking each person to repeat Step 2 after the group discussion; and
7) creating a group composite and summarizing major differences (if any).

To reiterate, this step serves two purposes. First, it determines which environmental assets to measure and their relative importance in assessing nutrient water quality. The advisory group's values and attributes can be directly input to the nutrient water quality assessment. Secondly, the stakeholders' relative value or weights of the environmental attributes emerge. Later, in ranking the nutrient water quality outcomes, the group's relative weights are used in determining the expected value of alternative SSNC.

Some facilitators conduct "issue rankings" to gauge the community's general environmental values and concerns. For example, the group may identify algal blooms or fish kills as general concerns. Although informative, issue ranking often fails to link nutrient water quality to measurable indicators that people care about, and it rarely promotes the formulation of relative values.

The next step in the stakeholder process is to determine relationships between stakeholder nutrient values and concerns, and nutrient and biological response variables. One approach for doing this would be to conduct a waterbody survey using the stakeholder advisory group in which perceived water quality would be evaluated by the stakeholders and compared with concurrent measurements of nutrients and biological response variables. Such a survey would have to be done over an extended period so that variations in water quality and associated stakeholder assessments could be determined. The literature shows that perception of nutrient-related water quality differs regionally. One of the best examples is from Minnesota, where large regional differences occur in the level of P, a surrogate of lake trophic state that is considered acceptable for drinking water supply and primary contact recreation (Table A-2). Perception regarding the level of algal biomass that constitutes a nuisance condition (Table A-3) also varies across the region. In Oregon, Chl *a* levels > 15 µg/L are considered to impair the beneficial uses of natural lakes, reservoirs, rivers, and estuaries, whereas in North Carolina, the Chl *a* criterion is 40 µg/L. Chl *a* criteria for Ontario lakes are quite restrictive relative to other regions in North America (Table A-4). These differences reflect regional water quality objectives and related uses that are attainable within the locale. A hypothetical example of such a survey is presented in

Tables A-5 through A-7. The results of such a survey would be combined with Tier 1, 2, and/or 3 SSNC evaluations to help select the most appropriate SSNC for the waterbody.

Table A-2. Lake Observer Survey

A.	Please circle the one number that best describes the physical condition of the lake water today.
1.	Crystal clear water.
2.	Not quite crystal clear, a little algae visible.
3.	Definite algal green, yellow, or brown color apparent.
4.	High algal levels with limited clarity and/or mild odor apparent.
5.	Severely high algal levels with one or more of the following: massive floating scums on lake or washed up on shore, strong foul odor, or fish kill.
B.	**Please circle the one number that best describes your opinion on how suitable the lake water is for recreation and aesthetic enjoyment today.**
1.	Beautiful, could not be any nicer.
2.	Very minor aesthetic problems, excellent for swimming, boating, enjoyment.
3.	Swimming and aesthetic enjoyment slightly impaired because of algal levels.
4.	Desire to swim and level of enjoyment of the lake substantially reduced because of algal levels (would not swim, but boating is okay).
5.	Swimming and aesthetic enjoyment of the lake nearly impossible because of algal levels.

Table A-3. Most Sensitive Lake Uses by Ecoregion and Corresponding Phosphorus Criteria for Minnesota (Heiskary and Walker 1988)

Ecoregion	Most Sensitive Uses	P Criteria (μg/L)
Northern Lakes and Forests	Drinking water supply	<15
	Cold water fishery	<15
	Primary contact recreation and aesthetics	<30
North Central Hardwood Forests	Drinking water supply	<30
	Primary contact recreation and aesthetics	<40
Western Corn Belt Plains	Drinking water supply	<40
	Primary contact recreation and aesthetics (full support) (partial support)	 <40 <90
Northern Glaciated Plains	Recreation and aesthetics (partial support)	<90

Table A-4. Aesthetic or Use Impairment Classification Systems Based on Chl *a* or Water Transparency (Heiskary and Walker 1988)

Author/Location	Chl *a* (μg/L)	Rating
Walmsley (1984) South African Reservoir	0–10 10–20 20–30 >30	No Problems Scums Evident Nuisance Severe Nuisance
Barica (1975) Canadian Prairie Ponds	0–25 25–100 100–200	Clear, No Blooms Moderate Blooms Dense Colonies and Scums
McGhee (1983) North Carolina	>15 >40	Unsuitable for Trout Severe Nuisance
Lillie and Mason (1983) Wisconsin	<1 1–5 5–10 10–15 15–30 >30	Excellent Very Good Good Fair Poor Very Poor

Table A-5. Potential Lake Uses Identified by Chlorophyll Levels for Lakes in Southern Ontario (Dillon and Rigler 1975)

Chl (μg/L)	Use Category
2	Body contact recreation and cold water fishery.
5	Water recreation where a cold water fishery is not imperative.
10	Body contact recreation of little importance, emphasis is on warm water fish.
25	Suitable for warm water fishery.

Table A-6. Hypothetical Lake Survey Part A

A. Describe condition of lake today		TN (ppb)	TP (ppb)	Secchi Depth (ft)	Chl *a* (ppb)
1	Crystal clear	30	15	10	3
2	Some algae visible	75	25	7	7
3	Green color	150	45	4	12
4	Limited clarity, mild odor	250	80	2	22
5	Floating scums, foul odor, dead fish	300	120	0.5	35

Table A-7. Hypothetical Lake Survey Part B

B. Describe suitability of lake for recreation		TN (ppb)	TP (ppb)	Secchi Depth (ft)	Chl *a* (ppb)
1	Beautiful, couldn't be nicer	30	15	10	3
2	Minor aesthetic problems, excellent for enjoyment	75	25	7	7
3	Swimming and aesthetics slightly impaired by algae	150	45	4	12
4	Swimming & aesthetics severely impaired by algae	250	80	2	22
5	Swimming & other recreation nearly impossible	300	120	0.5	35

REFERENCES

Allen, T.F.H. and T.B. Starr. 1982. Hierarchy–Perspectives for Ecological Complexity. University of Chicago Press, Chicago, Illinois.

Bachmann, R.W., B.L. Jones, D.D. Fox, M. Hoyer, L.A. Bull, and D.E. Canfield, Jr. 1996. Relations between trophic state indicators and fish in Florida (U.S.A.) lakes. *Can. J. Fish. Aquat. Sci.* 53:842-855.

Barbiero, R.P. and J. Kann. 1994. The importance of benthic recruitment to the population of development of *Aphanizomenon flos-aquae* and internal loading in a shallow lake. *J. Plankton Res.* 16:1581-1588.

Barbiero, R.P. and E.B. Welch. 1992. Contribution of benthic blue-green algal recruitment to lake population and phosphorus translocation. *Freshwat. Biol.* 27:249-260.

Barica, J. 1975. Summerkill risk in prairie ponds and possibilities of its prediction. *Can. J. Fish. Aquat. Sci.* 32:1283-1288.

Bartell, S.M. 1978. Size-selective Planktivory and Phosphorus Cycling in Pelagic Ecosystems. Ph.D. Dissertation, The University of Wisconsin, Madison, Wisconsin.

Bartell, S.M. 1994. Estimating sublethal toxic effects and ecological risks in coastal ecosystems, pp. 211-214. In: Dyer, K.R., and R.J. Orth, Eds.. Changes in Fluxes in Estuaries, ECSA22/ERF Symposium, Olsen & Olsen, Denmark.

Bartell, S.M., D. Breitburg, S. Brandt, D. Lipton, W. Lung, and D. Weller. An integrated modeling and mesocosm framework for assessing ecological risks in the Patuxent River watershed. *Estuaries* (in review).

Bartell, S.M., A.L. Brenkert, R.V. O'Neill, and R.H. Gardner. 1989. Temporal variation in regulation of production in a pelagic food web model, 101-118. In: Carpenter, S.R., Ed. Complex Interactions in Lake Communities. Springer-Verlag, New York.

Bartell, S.M., W.G. Cale, R.V. O'Neill, and R.H. Gardner. 1988. Aggregation error: research objectives and relevant model structure. *Ecol. Model.* 41:157-168.

Bartell, S.M., C.M. Campbell, K.R. Lovelock, S.K. Nair, and J.L. Shaw. 2000. Characterizing aquatic ecological risks from pesticides using a diquat dibromide case study III. Ecological process models. *Environ. Toxicol. Chem.* 19:1441-1453.

Bartell, S.M., R.H. Gardner, and R.V. O'Neill. 1992. Ecological Risk Estimation. Lewis Publishers, Inc., Chelsea, Michigan.

Bartell, S.M., G. Lefebvre, G. Kaminski, M. Carreau, and K.R. Campbell. 1999. An ecosystem model for assessing ecological risks in Québec rivers, lakes, and reservoirs. *Ecol. Model.* 124:43-67.

Bricker, S.B., C.G. Cement, D.E. Pirhalla, S.P. Orlando, and G.G.G. Farrow, D.G. 1999. National Estuarine Eutrophication Assessment: Effects of Nutrient Enrichment in the Nation's Estuaries. Special Projects Office and the National Centers for Coastal Ocean Science, National Ocean Service. National Oceanic and Atmospheric Administration. Silver Spring, Maryland.

Bostrom, B.J.M., S. Andersen, S. Fleischer, and M. Jansson. 1988. Exchange of P across the sediment water interface. *Hydrobiologia* 170:229-244.

Brown, L.C. and T.O. Barnwell, Jr. 1987. The Enhanced Stream Water Quality Models QUAL2E and QUAL2E-UNCAS: Documentation and User Manual. EPA/600/3-87/007, U.S. EPA, Athens, Georgia.

Buzzelli, C.P., R.L. Wetzel, and M.B. Meyers. 1999. A linked physical and biological framework to assess biogeochemical dynamics in a shallow estuarine ecosystem. *Estuarine Coastal and Shelf Science* 49:829-851.

Cale, W.G., R.V. O'Neill, and R.H. Gardner. 1983. Aggregation error in nonlinear ecosystem models. *J. Theor. Biol.* 100:539-550.

Campbell, K.R. and S.M. Bartell. 1998. Ecological models and ecological risk assessment. 69-100. In: Newman, M.C. and C.L. Trojan, Eds. Risk Assessment: Logic and Measurement. Ann Arbor Press. Chelsea, Michigan.

Canfield, D.E., Jr. and M.V. Hoyer. 1988. Regional geology and the chemical and trophic state characteristics of Florida lakes. *Lake Reserv. Manage.* 4:21-31.

Carlson, R.E. 1977. A trophic state index for lakes. *Limnol. Oceanogr.* 22:363-369.

Carlson, R.E. and J. Simpson. 1996. A Coordinator's Guide to Volunteer Lake Monitoring Methods. North American Lake Management Society, Madison, Wisconsin.

Carpenter, S.R. and J.F. Kitchell, Eds. 1993. The Trophic Cascade in Lakes. Cambridge University Press, Cambridge, England.

Cerco, C.F. and T. Cole. 1993. Three-dimensional eutrophication model of Chesapeake Bay. *J. Env. Eng.* 119:1006-1025.

Chang, J. and E.J. Carpenter. 1985. Blooms of the dinoflagellate *Gryodinium aureolum* in a Long Island estuary: Box model analysis of bloom maintenance. *Mar. Biol.* 89:83-93.

DeAngelis, D.L., S.M. Bartell, and A.L. Brenkert. 1989. Effects of nutrient cycling and food chain length on resilience. *Am. Nat*. 134:788-805.

Deegan, L.A., J.T. Finn, C.S. Hopkinson, A.E. Giblin, B.J. Peterson, B. Fry, and J.E. Hobbie. 1994. Flow model analysis of the effects of organic matter-nutrient interactions on estuarine trophic dynamics, 273-281. In: Dyer, K.R, and R.J. Orth, Eds. Changes in Fluxes in Estuaries. Olsen & Olsen, Fredensborg, Denmark.

Di Toro, D.M., D.J. O'Connor, and R.V. Thomann. 1971. A dynamic model of the phytoplankton population in the Sacramento-San Joaquin delta. *Am. Chem. Soc*. 106:131-180.

Di Toro, D.M., D.J. O'Connor, R.V. Thomann, and J.L. Mancini. 1975. Phytoplankton-zooplankton-nutrient interaction model for western Lake Erie, 424-474. In: Patten, B.C., Ed. Systems Analysis and Simulation in Ecology, Volume IV. Academic Press, New York.

Dillon, P.J. and G.H. Rigler. 1975. A simple method for predicting the capacity of a lake for development based on lake trophic status. *J. Fish. Res. Bd. Can*. 32:1519-1531.

Downing, J.A. and E. McCauley. 1992. The nitrogen:phosphorus relationship in lakes. *Limnol. Oceanogr*. 37:936-945.

Duarte, C.M. and J. Kalff. 1989. The influence of catchment geology and lake depth on phytoplankton biomass. *Arch. Hydrobiol*. 115:27-40.

Eldridge, P.M. and M.E. Sieracki. 1993. Biological and hydrodynamic regulation of the microbial food web in a periodically mixed estuary. *Limnol. Oceanogr*. 38:1666-1679.

Finn, J.T. 1976. Measures of ecosystem structure and function derived from analysis of flows. *J. Theor. Biol.* 56:363-380.

Fischoff, B. 1990. Understanding long-term environmental risks. *J. Risk Uncertainty*. 3:315-330.

Fitz, H.C., E.B. DeBellevue, R. Costanza, R. Boumans, T. Maxwell, L. Wainger, and F.H. Sklar. 1996. Development of a general ecosystem model for a range of scales and ecosystems. *Ecol. Model*. 88:263-295.

Fong, P., M.E. Jacobson, M.C. Mescher, D. Lirman, and M.C. Harwell. 1997. Investigating the management potential of a seagrass model through sensitivity analysis and experiments. *Ecol. Applic*. 7:300-315.

Forsberg, C. and S.O. Ryding. 1980. Eutrophication parameters and trophic state indices in 30 Swedish waste-receiving lakes. *Arch Hydrobiol.* 89:189-207.

Gallegos, C.L., T.E. Jordan, and D.L. Correll. 1992. Event-scale response of phytoplankton to watershed inputs in a subestuary: timing, magnitude, and location of blooms. *Limnology and Oceanography* 37:813-828.

Gardner, R.H., R.V. O'Neill, J.B. Mankin, and J.H. Carney. 1981. A comparison of sensitivity and error analysis based on a stream ecosystem model. *Ecol. Model.* 12:173-190.

Gerritsen, J., A.F. Holland, and D.E. Irvine. 1994. Suspension-feeding bivalves and the fate of primary production: an estuarine model applied to Chesapeake Bay. *Estuaries* 17:403-416.

Gordon, D.C., P.D. Keizer, D.R. Gaborn, P. Schwinghamer, and W.L. Silvert. 1986. Adventures in holistic ecosystem modeling: The Cumberland Basin ecosystem model. *Neth. J. Sea Res.* 20:325-335.

Gregory, R. and S. Lichtenstein. 1993. Valuing environmental resources: A constructive approach. *J. Risk Uncertainty.*7:177-197.

Ground, T.A. and A.W. Groeger. 1994. Chemical classification and trophic characteristics of Texas reservoirs. *Lake Reserv. Manage.* 10:189-201.

Hagy, J.D., W.R. Boynton, and L.P. Sanford. 2000. Estimation of net physical transport and hydraulic residence times for a coastal plain estuary using box models. *Estuaries* 23:328-340

Head, R.M., R.I. Jones, and A.E. Bailey-Watts. 1999. An assessment of the influence of recruitment from the sediment on the development of planktonic populations of cyanobacteria in a temperate mesotrophic lake. *Freshwat. Biol.* 41:759-769.

Heiskary, S.A. and W.W. Walker, Jr. 1988. Developing phosphorus criteria for Minnesota Lakes. *Lake Reserv. Manage.* 4:1-9.

Heymans, J.J. and D. Baird. 1995. Energy flow in the Kromme Estuarine Ecosystem, St. Francis Bay, South Africa. *Est. Coast Shelf Sci.* 41:39-59.

Hopkinson, Jr., C.S. and J.W. Day, Jr. 1977. Chapter 10: A model of the Barataria Bay salt marsh ecosystem. In: Ecosystem Modeling in Theory and Practice. C.A.S. Hall and J.W. Day, Jr., Eds. John Wiley & Sons,

Howarth, R.W., A. Giblin, B.H. Peterson, and G.W. Luther III. 1983. Reduced sulfur compounds in the pore waters of a New England salt marsh. *Environ. Biogeochem. Ecol. Bull.* 35:135-152.

Jones, J.R. and R.W. Bachmann. 1978. Trophic status of Iowa lakes in relation to origin and glacial geology. *Hydrobiologia* 57:267-273.

Jones, J.R. and M.F. Knowlton. 1993. Limnology of Missouri reservoirs: An analysis of regional patterns. *Lake Reserv. Manage.* 8:17-30.

Jones, J.R., M.F. Knowlton, and M.S. Kaiser. 1998. Effects of aggregation on chlorophyll-phosphorus relations in Missouri reservoirs. *Lake Reserv. Manage*. 14:1-9.

Jones, J.R. and J.T. Novak, 1981. Limnological characteristics of Lake of the Ozarks, Missouri. *Verh. Int. Ver. Theor. Agnew. Limnol*. 21:919-925.

Kelly, R.A. and W.O. Spofford, Jr. 1977. Chapter 18: Application of an ecosystem model to water quality management: The Delaware estuary. In: Hall, C.A.S., and J.W. Day, Jr., Eds. Ecosystem Modeling in Theory and Practice. John Wiley & Sons.

Kitchell, J.F., R.V. O'Neill, D. Webb., G. Gallepp, S.M. Bartell, J.F. Koonce, and B.S. Ausmus. 1979. Consumer regulation of nutrient cycling. *Bio. Science* 29:28-34.

Knowlton, M.F., M.V. Hoyer, and J.R. Jones. 1984. Sources of variability in phosphorus and chlorophyll and their effects on use of lake survey data. *Water Resour. Bull*. 20:397-407.

Kou, A.Y. and K. Park. 1995. A framework for coupling shoals and shallow embayments with main channels in a numerical modeling of coastal plain estuaries. *Estuaries* 18:341-350

Kremer, J.N. and S.W. Nixon. 1978. A Coastal Marine Ecosystem. Simulation and Analysis. Springer-Verlag, New York.

Lillie, R.A. and J.W. Mason. 1983. Limnological Characteristics of Wisconsin Lakes. Tech. Bull No. 138., Dept. Nat. Resour., Madison, Wisconsin.

Lucas, L.V., J.E. Cloern, J.R. Koseff, S.G. Monismith, and J.K. Thompson. 1998. Does the Sverdrup critical depth model explain bloom dynamics in estuaries? *J. Mar. Res*. 56:375-415.

MacCormick, A.J.A., O.L. Loucks, J.F. Koonce, J.F. Kitchell, and P.R. Weiler. 1975. An ecosystem model for the pelagic zone of a lake. *Environ. Syst. Sci.* 2:339-382.

Madden, C.J. and W.M. Kemp. 1996. Ecosystem model of an estuarine submersed plant community: calibration and simulation of eutrophication responses. *Estuaries* 19:457-474.

Marsden, M.W. 1989. Lake restoration by reducing external phosphorus loading: The influence of sediment phosphorus release. *Freshwat. Biol*. 21:139-162.

McGhee, R.F. 1983. Experience in developing a chlorophyll *a* standard in the Southeast to protect lakes, reservoirs and estuaries. Lake Restoration Protection and Management, Proc. 2nd Annu. Conf. N. Am. Lake Manage. Soc. EPA 440/5-83-001. U.S. Environmental Protection Agency, Washington, D.C.

McIntosh, R.P. 1985. The Background of Ecology Concept and Theory. Cambridge University Press, Cambridge, U.K.

Morris, D.P. and W.M. Lewis, Jr. 1988. Phytoplanton nutrient limitation in Colorado mountain lakes.

National Academy of Sciences. 1969. Eutrophication: Causes, consequences, correctives. National Academy of Sciences, Washington, D.C.

Newbold, J.D., J.W. Elwood, R.V. O'Neill, and W. Van Winkle. 1981. Measuring nutrient spiraling in streams. *Can. J. Fish. Aquat. Sci.* 38:860-863.

Nixon, S.W. and J.N. Kremer. 1977. Narragansett Bay—the development of a composite simulation model for a New England estuary. In: Hall C.A.S., and J.W. Day, Jr., Eds. Ecosystem Modeling in Theory and Practice. John Wiley & Sons.

Nurnberg, G.K. 1996. Trophic state of clear and colored, soft- and hardwater lakes with special consideration of nutrients, anoxia, phytoplankton and fish. *Lake Reserv. Manage.* 12:432-447.

O'Neill, R.V. and B. Rust. 1979. Aggregation error in ecological models. *Eco. Model.* 7:91-105.

Odum, E.P. 1971. Fundamentals of Ecology. W.B. Saunders, Philadelphia, Pennsylvania.

Odum, H.T. 1983. Systems Ecology: An Introduction. John Wiley & Sons, N.Y.

Oglesby, R.T. 1977. Relationship of fish yield to lake phytoplankton standing crop, production, and morphoedaphic factors. *J. Fish. Res. Bd. Can.* 34:2271-2279.

Omernik, J.M., C.M. Rohm, R.A. Lillie, and N. Mesner. 1991. Usefulness of natural regions for lake management: Analysis of variation among lakes in northwestern Wisconsin, U.S.A. *Environ. Manage*. 15:281-293.

Osgood, R.A. 1988. Lake mixis and internal phosphorus dynamics. *Arch. Hydrobiol*. 113:629-638.

Park, R.A. 1998. AQUATOX for Windows: A Modular Toxic Effects Model for Aquatic Ecosystems. Contract No. 68-C4-0051, 3-13. U.S. Environmental Protection Agency, Washington, D.C.

Park, R.A., R.V. O'Neill, J.A. Bloomfield, H.H. Shugart, Jr., R.S. Booth, R.A. Goldstein, J.B. Mankin, J.F. Koonce, D. Scavia, M.S. Adams, L.S. Clesceri, E.M. Colon, E.H. Dettman, J.A. Hoopes, D.D. Huff, S. Katz, J.F. Kitchell, R.C. Kohberger, E.J. LaRow, D.C. McNaught, J.L. Peterson, J.E. Titus, P.W. Weiler, J.W. Wilkinson, and C.S. Zahorcak. 1974. A generalized model for simulating lake ecosystems. *Simulation* 21:33-50.

Pastorok, R.A., S.M. Bartell, S. Ferson, and L.R. Ginzburg. 2002. Ecological Modeling in Risk Assessment. CRC Press, Boca Raton, Florida.

Patten, B.C. 1971. Systems Analysis and Simulation in Ecology. Volume 1. Academic Press, New York, N.Y.

Patten, B.C., D.A. Egloff, and T.H. Richardson. 1975. Total ecosystem model for a cove in Lake Texoma, 205-421. In: Patten, B.C., Ed. Systems Analysis and Simulation in Ecology, Volume IV. Academic Press, New York.

Paul, R.W. 2001. Geographical signatures of middle Atlantic estuaries: historical layers. *Estuaries* 24:151-166.

Peterson, D.L. and V.T. Parker. 1998. Dimensions of scale in ecology, resource management, and society, 499-522. In: Peterson, D.L. and V.T. Parker, Eds. Ecological Scale: Theory and Applications. Columbia University Press, New York.

Radford, P.J. 1979. Some aspects of an estuarine ecosystem model-GEMBASE, 301-322. In: S.E. Jorgensen, Ed. State-of-the-art in Ecological Modeling, Volume 7. International Society for Ecological Modeling, Copenhagen.

Rast, W. and G.F. Lee. 1978. Summary Analysis of the North American Project (U.S. portion) OECD Eutrophication Project: Nutrient Loading-Lake Response Relationships and Trophic State Indices. Corvallis, Oregon, U.S. EPA, Corvallis Environmental Research Lab.

Rawson, D.S. 1939. Some physical and chemical factors in the metabolism of lakesCproblems of lake biology. *Amer. Assoc. Adv. Sci.* 10:1-26.

Reckhow, K.H. 1988. Empirical models for trophic state in southeastern U.S. lakes and reservoirs. *Wat. Resour. Bul.* 24:723-734.

Schindler, D.W. 1977. Evolution of phosphorus limitation in lakes. *Science* 195:260-262.

Schindler, D.W. 1990. Experimental perturbation of whole lakes as tests of hypotheses concerning ecosystem structure and function. *Oikos* 57:25-41.

Smith, V.H. 1998. Cultural eutrophication of inland, estuarine, and coastal waters. In: M.L Pace and P.M. Groffman, Eds. Successes, Limitations and Frontiers in Ecosystem Ecology, Springer-Verlag, New York.

Stockner, J.G. and K.G. Porter. 1989. Microbial food webs in freshwater planktonic ecosystems, 69-83. In: Carpenter, S.R., Ed. Complex Interactions in Lake Communities. Springer-Verlag, New York.

Swartzman, G. and S. Kaluzny. 1987. Ecological Simulation Primer. MacMillan Publishing Company, New York.

Thomann, R.V. and L.C. Linker. 1998. Contemporary issues in watershed and water quality modeling for eutrophication control. *Wat. Sci. Tech.* 37:93-102.

Tomovic, R. and W.J. Karplus. 1963. Sensitivity Analysis of Dynamic Systems. McGraw-Hill, New York.

U.S. Environmental Protection Agency. 1983. Technical Support Manual: Waterbody Surveys and Assessments for Conducting Use Attainability Analyses. U.S. Environmental Protection Agency, Criteria and Standards Division, Washington, D.C.

U.S. Environmental Protection Agency. 1995. QUAL2E—The Enhanced Stream Water Quality Model. EPA-823-B-95-003. U.S. Environmental Protection Agency, Environmental Research Laboratory, Office of Research and Development, Athens, Georgia.

U.S. Environmental Protection Agency. 1996. National Water Quality Inventory 1996 Report to Congress. U.S. Environmental Protection Agency, Office of Water, Washington, D.C.

U.S. Environmental Protection Agency. 1998. National Strategy for the Development of Regional Nutrients. EPA-822-R-98-002. U.S. Environmental Protection Agency, Office of Water, Washington, D.C.

U.S. Environmental Protection Agency. 2000a. Nutrient Criteria Technical Guidance Manual for Lakes and Reservoirs. EPA-822-B00-001. U.S. Environmental Protection Agency, Washington, D.C.

U.S. Environmental Protection Agency. 2000c. Ambient Water Quality Criteria Recommendations. Information supporting the Development of State and Tribal Nutrient Criteria. Rivers and Streams in Nutrient Ecoregion XIV. EPA 822-B-00-022. United States Environmental Protection Agency, Office of Water, Washington, D.C.

U.S. Environmental Protection Agency. 2000b. Nutrient Criteria Technical Guidance Manual for River and Streams. EPA-822-B-00-002. United States Environmental Protection Agency, Washington, D.C.

U.S. Environmental Protection Agency. 2001. Better Assessment Science Integrating Point and Nonpoint Sources: BASINS Version 3.0 User's Manual. EPA-823-B-01-001. U.S. Environmental Protection Agency, Office of Science and Technology, Washington, D.C.

Vannote, R.L., G.W. Minshall, K.W. Cummins, J.R. Sedell, and C.E. Cushing. 1980. The river continuum concept. *Can. J. Fish. Aquat. Sci.* 37:130-137.

Vollenweider, R.A. 1968. Scientific Fundamentals of Lake and Stream Eutrophication, with Particular Reference to Phosphorus and Nitrogen as Eutrophication Factors. Tech. Rep. DAS/DSI/68.27, OECD, Paris.

Vollenweider, R.A. 1976. Advances in defining loading levels for phosphorus in lake eutrophication. *Mem. 1st Ital. Idrabio*. 33:53-83.

Walker, W.W. 1979. Use of hypolimnetic oxygen depletion rate as a trophic state index for lakes. *Wat. Resour. Res*. 15:1463-1470.

Walmsley, R.D. 1984. A chlorophyll *a* trophic status classification system for South African impoundments. *J. Envir. Qual.* 13:97-104.

Ward, J.V. and J.A. Stanford. 1983. The serial discontinuity concept of lotic ecosystems, 29-42. In: Fontaine, T.D., and S.M. Bartell, Eds. Dynamics of Lotic Ecosystems. Ann Arbor Science, Ann Arbor, Michigan.

Warren-Hicks, W.J. and D.R.J Moore, Eds. 1998. Uncertainty Analysis in Ecological Risk Assessment. SETAC Press.

Water Environment Research Foundation. 1995. A Comprehensive UAA Technical Reference and User's Guide. Project 91-NPS-1. Water Environment Research Foundation, Alexandria, Virginia.

Welch, E.B. and G.D. Cooke. 1995. Internal phosphorus loading in shallow lakes: Importance and control. *Lake Reserv. Manage.* 11:273-281.

Wetzel, R. 1983. Limnology, 2nd edition, W.B. Saunders, Philadelphia, Pennsylvania.

Yurk, J.J. and J.J. Ney. 1989. Phosphorus-fish community biomass relationships in southern Appalachian reservoirs: Can lakes be too clean for fish? *Lake Reserv. Manage.* 5:83-90.

Zeigler, B.P. 1976. The aggregation problem, 299-311. In, B.C. Patten, Ed. Systems Analysis and Simulation in Ecology. Volume 4. Academic Press, New York.

WERF SUBSCRIBERS

WASTEWATER UTILITY

Alabama
Montgomery Water Works & Sanitary Sewer Board

Alaska
Anchorage Water & Wastewater Utility

Arizona
Gila Resources
Glendale, City of, Utilities Department
Mesa, City of
Peoria, City of
Pima County Wastewater Management

Arkansas
Little Rock Wastewater Utility

California
Calaveras County Water District
Central Contra Costa Sanitary District
Crestline Sanitation District
Delta Diablo Sanitation District
Dublin San Ramon Services District
East Bay Dischargers Authority
East Bay Municipal Utility District
Eastern Municipal Water District
El Dorado Irrigation District
Fairfield-Suisun Sewer District
Fort Lauderdale, City of
Irvine Ranch Water District
Las Virgenes Municipal Water District
Livermore, City of
Lodi, City of
Los Angeles, City of
Los Angeles County, Sanitation Districts of
Napa Sanitation District
Orange County Sanitation District
Palo Alto, City of
Riverside, City of
Sacramento Regional County Sanitation District
San Diego Metropolitan Wastewater Department, City of
San Francisco, City & County of
San Jose, City of
Santa Barbara, City of
Santa Cruz, City of
Santa Rosa, City of
South Bayside System Authority
South Coast Water District
South Orange County Wastewater Authority
Sunnyvale, City of
Union Sanitary District
West Valley Sanitation District

Colorado
Boulder, City of
Colorado Springs Utilities
Greeley, City of
Littleton/Englewood Water Pollution Control Plant
Metro Wastewater Reclamation District, Denver

Connecticut
The Mattabassett District
New Haven, City of, WPCA

District of Columbia
District of Columbia Water & Sewer Authority

Florida
Broward, County of
Fort Lauderdale, City of
Gainesville Regional Utilities
JEA
Miami-Dade Water & Sewer Authority
Orange County Utilities Department
Reedy Creek Improvement District
Seminole County Environmental Services
St. Petersburg, City of
Stuart Public Utilities
Tallahassee, City of
Tampa, City of
Toho Water Authority
West Palm Beach, City of

Georgia
Atlanta Department of Watershed Management
Augusta, City of
Clayton County Water Authority
Cobb County Water System
Columbus Water Works
Fulton County
Gwinnett County Department of Public Utilities
Savannah, City of

Hawaii
Honolulu, City & County of

Idaho
Boise, City of

Illinois
American Bottoms Wastewater Treatment Plant
Greater Peoria Sanitary District
Kankakee River Metropolitan Agency
Metropolitan Water Reclamation District of Greater Chicago
Wheaton Sanitary District

Iowa
Ames, City of
Des Moines Metro Wastewater Reclamation Authority
Iowa City

Kansas
Johnson County Unified Wastewater Districts
Unified Government of Wyandotte County/ Kansas City, City of

Kentucky
Louisville & Jefferson County Metropolitan Sewer District

Louisiana
Sewerage & Water Board of New Orleans

Maine
Bangor, City of
Portland Water District

Maryland
Anne Arundel County Bureau of Utility Operations
Howard County Department of Public Works
Washington Suburban Sanitary Commission

Massachusetts
Boston Water & Sewer Commission
Upper Blackstone Water Pollution Abatement District

Michigan
Ann Arbor, City of
Detroit, City of
Holland Board of Public Works
Lansing, City of
Saginaw, City of
Wayne County Department of Environment
Wyoming, City of

Minnesota
Rochester, City of
Western Lake Superior Sanitary District

Missouri
Independence, City of
Kansas City Missouri Water Services Department
Little Blue Valley Sewer District
Metropolitan St. Louis Sewer District

Nebraska
Lincoln Wastewater System

Nevada
Henderson, City of

New Jersey
Bergen County Utilities Authority
Passaic Valley Sewerage Commissioners

New York
New York City Department of Environmental Protection

North Carolina
Charlotte/Mecklenburg Utilities
Durham, City of
Metropolitan Sewerage District of Buncombe County
Orange Water & Sewer Authority

Ohio
Akron, City of
Butler County Department of Environmental Services
Columbus, City of
Metropolitan Sewer District of Greater Cincinnati
Northeast Ohio Regional Sewer District
Summit, County of

Oklahoma
Trinity River Authority
Tulsa, City of

Oregon
Clean Water Services
Eugene/Springfield Water Pollution Control
Water Environment Services

Pennsylvania
Philadelphia, City of
University Area Joint Authority

South Carolina
Charleston Commissioners of Public Works
Mount Pleasant Waterworks & Sewer Commission
Spartanburg Sanitary Sewer District

Tennessee
Cleveland, City of
Knoxville Utilities Board
Murfreesboro Water & Sewer Department
Nashville Metro Water Services

Texas
Austin, City of
Dallas Water Utilities
Denton, City of
El Paso Water Utilities
Fort Worth, City of
Gulf Coast Waste Disposal Authority
Houston, City of
San Antonio Water System
Trinity River Authority

Utah
Salt Lake City Corporation

Virginia
Alexandria Sanitation Authority
Arlington, County of
Fairfax County
Hampton Roads Sanitation District
Henrico, County of

Hopewell Regional Wastewater Treatment Facility
Loudoun County Sanitation Authority
Lynchburg Regional WWTP
Prince William County Service Authority
Richmond, City of
Rivanna Water & Sewer Authority

Washington
Bellevue Utilities Department
Edmonds, City of
Everett, City of
King County Department of Natural Resources
Seattle Public Utilities
Sunnyside, Port of
Yakima, City of

Wisconsin
Green Bay Metro Sewerage District
Madison Metropolitan Sewerage District
Milwaukee Metropolitan Sewerage District
Racine, City of
Sheboygan Regional Wastewater Treatment
Wausau Water Works

Australia
South Australian Water Corporation
Sydney Water Corporation
Water Corporation of Western Australia

Canada
Greater Vancouver Regional District
Regina, City of, Saskatchewan
Toronto, City of, Ontario
Winnipeg, City of, Manitoba

Mexico
Servicios de Agua y Drenaje de Monterrey, I.P.D.

New Zealand
Watercare Services Limited

United Kingdom
Yorkshire Water Services Limited

STORMWATER UTILITY

California
Los Angeles, City of, Department of Public Works
Monterey, City of
San Francisco, City & County of
Santa Rosa, City of

Colorado
Boulder, City of

Georgia
Griffin, City of

Iowa
Cedar Rapids Wastewater Facility
Des Moines Metro Wastewater Reclamation Authority

Kansas
Overland Park, City of

Maine
Portland Water District

North Carolina
Charlotte, City of, Stormwater Services

Pennsylvania
Philadelphia, City of

Tennessee
Chattanooga Stormwater Management

Washington
Seattle Public Utilities

STATE

Arkansas Department of Environmental Quality
Fresno Department of Public Utilities
Kansas Department of Health & Environment
Urban Drainage & Flood Control District, CO

CORPORATE

ADS Environmental Services
The ADVENT Group Inc.
Alan Plummer & Associates
Alden Research Laboratory Inc.
Aqua-Aerobic Systems Inc.
Aquateam–Norwegian Water Technology Centre A/S
ARCADIS
Associated Engineering
BaySaver Inc.
Black & Veatch
Blasland, Bouck & Lee Inc.
Boyle Engineering Corporation
Brown & Caldwell
Burns & McDonnell
CABE Associates Inc.
The Cadmus Group
Camp Dresser & McKee Inc.
Carollo Engineers Inc.
Carpenter Environmental Associates Inc.
CDS Technologies Inc.
Chemtrac Systems Inc.
CH2M HILL
Damon S. Williams Associates, LLC
David L. Sheridan, P.C.
Dewling Associates Inc.
Earth Tech Inc.
Ecovation
EMA Inc.
The Eshelman Company Inc.
EWT Holdings Corporation
Fay, Spofford, & Thorndike Inc.
Freese & Nichols Inc.
ftn Associates Inc.
Fuss & O'Neill Inc.
Gannett Fleming Inc.
GE Betz
Greeley and Hansen LLC
Hazen & Sawyer, P.C.
HDR Engineering Inc.
HNTB Corporation
HydroQual Inc.
Infilco Degremont Inc.
Ingersoll-Rand Energy Systems
Insituform Technologies Inc.
Institute for Environmental Technology & Industry, Korea
Jacobson Helgoth Consultants Inc.
Jason Consultants LLC Inc.
Jordan, Jones, & Goulding Inc.
KCI Technologies Inc.
Kelly & Weaver, P.C.
Kennedy/Jenks Consultants
Komline Sanderson Engineering Corporation
Lawler, Matusky & Skelly Engineers, LLP
LTI Inc.
Lombardo Associates Inc.
Malcolm Pirnie Inc.
McKim & Creed
MEC Analytical Systems Inc.
Metcalf & Eddy Inc.
MWH
New England Organics
O'Brien & Gere Engineers Inc.
Odor & Corrosion Technology Consultants Inc.
Oswald Green, LLC
PA Government Services Inc.
Parametrix Inc.
Parsons
Post, Buckley, Schuh & Jernigan
R&D Engineering/Conestoga Rover & Associates
RMC Inc.
R.M. Towill Corporation
Ross & Associates Ltd.
Royce Technologies
SAIC Maritime Technical Group
Short Elliott Hendrickson Inc.
Stantec Consulting Inc.
Stearns & Wheler, LLC
Stone Environmental Inc.
Stormwater Management Inc.
Stratus Consulting Inc.
Synagro Technologies Inc.
Tetra Tech Inc.
Trojan Technologies Inc.
URS Corporation
USFilter
Veola Water North America
Wade-Trim Inc.
Westin Engineering Inc.
Weston Solutions Inc.
Woodard & Curran
WRc/D&B, LLC
WWETCO, LLC
Zenon Environmental Inc.
Zoeller Pump Company

INDUSTRY

American Electric Power
ChevronTexaco Energy Research & Technology Company
The Coca-Cola Company
Dow Chemical Company
DuPont Company
Eastman Chemical company
Eastman Kodak Company
Eli Lilly & Company
Merck & Company Inc.
ONDEO Services
Procter & Gamble Company
PSEG Services Corporation
RWE Thames Water Plc
Severn Trent Services Inc.
United Water Services LLC

Note: List as of 2/1/05

WERF Product Order Form

As a benefit of joining the Water Environment Research Foundation, subscribers are entitled to receive one complimentary copy of all final reports and other products. Additional copies are available at cost (usually $10). To order your complimentary copy of a report, please write "free" in the unit price column. WERF keeps track of all orders. If the charge differs from what is shown here, we will call to confirm the total before processing.

Name | Title

Organization

Address

City | State | Zip Code | Country

Phone | Fax | Email

Stock #	Product	Quantity	Unit Price	Total
			Postage & Handling	
			VA Residents Add 4.5% Sales Tax	
			Canadian Residents Add 7% GST	
			TOTAL	

Method of Payment:

(All orders must be prepaid.)

❑ Check or Money Order Enclosed

❑ Visa ❑ Mastercard ❑ American Express

Account No. | Exp. Date

Signature

Shipping & Handling:

Amount of Order	United States	Canada & Mexico	All Others
Up to but not more than:	Add:	Add:	Add:
$20.00	$5.00*	$8.00	50% of amount
30.00	5.50	8.00	40% of amount
40.00	6.00	8.00	
50.00	6.50	14.00	
60.00	7.00	14.00	
80.00	8.00	14.00	
100.00	10.00	21.00	
150.00	12.50	28.00	
200.00	15.00	35.00	
More than $200.00	Add 20% of order	Add 20% of order	

*minimum amount for all orders

Note: Please make checks payable to the Water Environment Research Foundation.

To Order (Subscribers Only):

Log on to www.werf.org and click on the "Product Catalog."

Phone: (703) 684-2470
Fax: (703) 299-0742.

WERF
Attn: Subscriber Services
635 Slaters Lane
Alexandria, VA 22314-1177

To Order (Non-Subscribers):

Non-subscribers may be able to order WERF publications either through WEF (www.wef.org) or IWAP (www.iwapublishing.com). Visit WERF's website at www.werf.org for details.

IWA Publishing's authorised EU representative for General Product Safety Regulations is Diane D'Arras, 15 rue Duret, 75116 Paris, France, e-mail: safety@iwap.co.uk.

Printed and bound by CPI Group (UK) Ltd, Croydon, CR0 4YY
12/05/2026
02108881-0015